JANINE
M. BENYUS

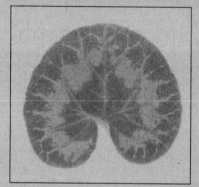

BIOMIMICRY

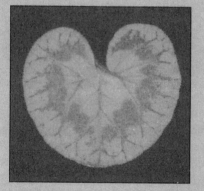

INNOVATION
INSPIRED
BY NATURE

HARPER **PERENNIAL**

HARPER ● PERENNIAL

HarperCollins books may be purchased for educational, business, or sales promotional use. For information please write: Special Markets Department, HarperCollins Publishers Inc., 10 East 53rd Street, New York, NY 10022.

First Quill edition published 1998.

Reissued in Perennial 2002.

Designed by Oksana Kushnir

Library of Congress Cataloging-in-Publication Data
Benyus, Janine M.
    Biomimicry : innovation inspired by nature / Janine M. Benyus.
      p. cm.
Includes bibliographical references (p.) and index.
    ISBN 0-688-16099-9 (pbk.)
     1. Technological innovations. 2. Human ecology. 3. Nature. I. Title.
    T173.8.B45 1997
    600—dc21   96-52336
                                      CIP

ISBN 0-06-053322-6 (reissue) ISBN 978-0-06-053322-9

06 07 08 RRD 20 19 18 17 16 15 14 13 12

FOR THE MENTORS

ON THE TANGLED BANK

ALSO BY JANINE M. BENYUS

Beastly Behaviors: A Watcher's Guide to How Animals Act and Why

Northwoods Wildlife: A Watcher's Guide to Habitats

Northwoods Wildlife—Knapsack Edition

The Field Guide to Wildlife Habitats of the Eastern United States

The Field Guide to Wildlife Habitats of the Western United States

ACKNOWLEDGMENTS

I wish to express my appreciation to all the biomimics I interviewed and especially to those kind enough to review a portion of the manuscript. The reviewers were: Dr. Wes Jackson, Dr. Jon Piper, and Dr. Marty Bender of The Land Institute; Dr. J. Devens Gust, Jr., Dr. Thomas Moore, Dr. Ana Moore, and Dr. Neal Woodbury of Arizona State University; Dr. Clement Furlong, University of Washington; Dr. Paul Calvert, University of Arizona; Dr. J. Herbert Waite, University of Delaware; Dr. Christopher Viney, Oxford University; Dr. David Kaplan, U.S. Army Research; Dr. Kenneth Glander, Duke University Primate Center; Dr. Richard Wrangham, Harvard University; Dr. Karen Strier, University of Wisconsin; Dr. Michael Conrad, Wayne State University; Dr. Braden Allenby and Dr. Thomas Graedel of AT&T; and Thomas Armstrong of Matfield Green, Kansas. I owe a special debt of gratitude to Dr. Christopher Viney, who critiqued the entire manuscript with a rare combination of enthusiasm and a fine-tooth comb.

I was fortunate to have a literary agent, Jeanne Hanson, and an editor, Toni Sciarra, who really understood this field-without-a-name and were biomimicry champions from the start. For transcribing my notes with a curious mind, I thank Nina Maclean. My flock of friends and family were tremendous, as always.

Many people shaped my understanding of this book, both while I was writing it and afterward. In particular, I thank Wes Jackson and Wendell Berry for recognizing themselves as biomimics years ago and thinking so clearly and carefully about what it all means. Emily Hunter, also of The Land Institute, was waiting in an eddy for me when I finished. With her help, I was able to reflect and recharge for the next phase.

Finally, I want to thank Laura Merrill, who, with patient ear and open heart, helped midwife the birth of biomimicry. Her otterlike joy and rock-steady support has meant the world.

CONTENTS

CHAPTER 1 **ECHOING NATURE**
WHY BIOMIMICRY NOW? 1

CHAPTER 2 **HOW WILL WE FEED OURSELVES?**
FARMING TO FIT THE LAND: GROWING FOOD
LIKE A PRAIRIE 11

CHAPTER 3 **HOW WILL WE HARNESS ENERGY?**
LIGHT INTO LIFE: GATHERING ENERGY LIKE
A LEAF 59

CHAPTER 4 **HOW WILL WE MAKE THINGS?**
FITTING FORM TO FUNCTION: WEAVING FIBERS
LIKE A SPIDER 95

CHAPTER 5 **HOW WILL WE HEAL OURSELVES?**
EXPERTS IN OUR MIDST: FINDING CURES LIKE
A CHIMP 146

CHAPTER 6 **HOW WILL WE STORE WHAT WE LEARN?**
DANCES WITH MOLECULES: COMPUTING LIKE
A CELL 185

CHAPTER 7 **HOW WILL WE CONDUCT BUSINESS?**
CLOSING THE LOOPS IN COMMERCE: RUNNING A
BUSINESS LIKE A REDWOOD FOREST 238

CHAPTER 8 **WHERE WILL WE GO FROM HERE?**
MAY WONDERS NEVER CEASE: TOWARD A
BIOMIMETIC FUTURE 285

BIO-INSPIRED READINGS 299
INDEX 301

BI-O-MIM-IC-RY

[From the Greek bios, life, and mimesis, imitation]

1. *Nature as model.* Biomimicry is a new science that studies nature's models and then imitates or takes inspiration from these designs and processes to solve human problems, e.g., a solar cell inspired by a leaf.

2. *Nature as measure.* Biomimicry uses an ecological standard to judge the "rightness" of our innovations. After 3.8 billion years of evolution, nature has learned: What works. What is appropriate. What lasts.

3. *Nature as mentor.* Biomimicry is a new way of viewing and valuing nature. It introduces an era based not on what we can *extract* from the natural world, but on what we can *learn* from it.

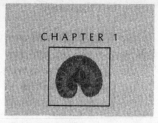

ECHOING NATURE

WHY BIOMIMICRY NOW?

We must draw our standards from the natural world. We must honor with the humility of the wise the bounds of that natural world and the mystery which lies beyond them, admitting that there is something in the order of being which evidently exceeds all our competence.
—VÁCLAV HAVEL, president of the Czech Republic

It's not ordinary for a bare-chested man wearing jaguar teeth and owl feathers to grace the pages of *The New Yorker*, but these are not ordinary times. While I was writing this book, Moi, an Huaorani Indian leader whose name means "dream," traveled to Washington, D.C., to defend his Amazonian homeland against oil drilling. He roared like a jaguar in the hearings, teaching a roomful of jaded staffers where real power comes from and what homeland actually means.

Meanwhile, in America's heartland, two books about aboriginal peoples were becoming word-of-mouth best-sellers, much to their publishers' surprise. Both were about urban Westerners whose lives are changed forever by the wise teachings of preindustrial societies.

What's going on here? My guess is that *Homo industrialis*, having reached the limits of nature's tolerance, is seeing his shadow on the wall, along with the shadows of rhinos, condors, manatees, lady's slippers, and other species he is taking down with him. Shaken by the sight, he, we, are hungry for instructions about how to live sanely and sustainably on the Earth.

The good news is that wisdom is widespread, not only in indigenous peoples but also in the species that have lived on Earth far longer than humans. If the age of the Earth were a calendar year and today were a breath before midnight on New Year's Eve, we showed up a scant fifteen minutes ago, and all of recorded history has blinked by in the last sixty seconds. Luckily for us, our planet-mates—the fantastic meshwork of plants, animals, and microbes—have been patiently perfecting their wares since March, an incredible 3.8 billion years since the first bacteria.

In that time, life has learned to fly, circumnavigate the globe, live in the depths of the ocean and atop the highest peaks, craft miracle materials, light up the night, lasso the sun's energy, and build a self-reflective brain. Collectively, organisms have managed to turn rock and sea into a life-friendly home, with steady temperatures and smoothly percolating cycles. In short, living things have done everything we want to do, without guzzling fossil fuel, polluting the planet, or mortgaging their future. What better models could there be?

ECHO-INVENTIONS

In these pages, you'll meet men and women who are exploring nature's masterpieces—photosynthesis, self-assembly, natural selection, self-sustaining ecosystems, eyes and ears and skin and shells, talking neurons, natural medicines, and more—and then copying these designs and manufacturing processes to solve our own problems. I call their quest *biomimicry*—the conscious emulation of life's genius. Innovation inspired by nature.

In a society accustomed to dominating or "improving" nature, this respectful imitation is a radically new approach, a revolution really. Unlike the Industrial Revolution, the Biomimicry Revolution introduces an era based not on what we can *extract* from nature, but on what we can *learn* from her.

As you will see, "doing it nature's way" has the potential to change the way we grow food, make materials, harness energy, heal ourselves, store information, and conduct business.

In a biomimetic world, we would manufacture the way animals and plants do, using sun and simple compounds to produce totally biodegradable fibers, ceramics, plastics, and chemicals. Our farms, modeled on prairies, would be self-fertilizing and pest-resistant. To find new drugs or crops, we would consult animals and insects that have used plants for millions of years to keep themselves healthy and nourished. Even computing would take its cue from nature, with software that "evolves" solutions, and hardware that uses the lock-and-key paradigm to compute by touch.

In each case, nature would provide the models: solar cells copied from leaves, steely fibers woven spider-style, shatterproof ceramics drawn from mother-of-pearl, cancer cures compliments of chimpanzees, perennial grains inspired by tallgrass, computers that signal like cells, and a closed-loop economy that takes its lessons from redwoods, coral reefs, and oak-hickory forests.

The biomimics are discovering what works in the natural world, and more important, what lasts. After 3.8 billion years of research and development, failures are fossils, and what surrounds us is the secret to survival. The more our world looks and functions like this natural world, the more likely we are to be accepted on this home that is ours, but not ours alone.

This, of course, is not news to the Huaorani Indians. Virtually all native cultures that have survived without fouling their nests have acknowledged that nature knows best, and have had the humility to ask the bears and wolves and ravens and redwoods for guidance. They can only wonder why we don't do the same. A few years ago, I began to wonder too. After three hundred years of Western Science, was there anyone in our tradition able to see what the Huaorani see?

HOW I FOUND THE BIOMIMICS

My own degree is in an applied science—forestry—complete with courses in botany, soils, water, wildlife, pathology, and tree growth. Especially tree growth. As I remember, cooperative relationships, self-regulating feedback cycles, and dense interconnectedness were not something we needed to know for the exam. In reductionist fashion, we studied each piece of the forest separately, rarely considering that a spruce-fir forest might add up to something more than the sum of its parts, or that wisdom might reside in the whole. There were no labs in listening to the land or in emulating the ways in which natural communities grew and prospered. We practiced a human-centered approach to management, assuming that nature's way of managing had nothing of value to teach us.

It wasn't until I started writing books on wildlife habitats and behavior that I began to see where the real lessons lie: in the exquisite ways that organisms are adapted to their places and to each other. This hand-in-glove harmony was a constant source of delight to me, as well as an object lesson. In seeing how seamlessly animals fit into their homes, I began to see how separate we managers had become from ours. Despite the fact that we face the same physical challenges that all living beings face—the struggle for food, water, space, and shelter in a finite habitat—we were trying to meet those challenges through human cleverness alone. The lessons inherent in the natural world, strategies sculpted and burnished over billions of years, remained scientific curiosities, divorced from the business of our lives.

But what if I went back to school now? Could I find any researchers who were consciously looking to organisms and ecosystems for inspiration about how to live lightly and ingeniously on the Earth? Could I work with inventors or engineers who were dipping into biology texts for ideas? Was there anyone, in this day and age, who regarded organisms and natural systems as the ultimate teachers?

Happily, I found not one but many biomimics. They are fascinating people, working at the edges of their disciplines, in the fertile crests between intellectual habitats. Where ecology meets agriculture, medicine, materials science, energy, computing, and commerce, they are learning that there is more to discover than to invent. They know that nature, imaginative by necessity, has already solved the problems we are struggling to solve. Our challenge is to take these time-tested ideas and echo them in our own lives.

Once I found the biomimics, I was thrilled, but surprised that there is no formal movement as yet, no think tanks or university degrees in biomimicry. This was strange, because whenever I mentioned what I was working on, people responded with a universal enthusiasm, a sort of relief upon hearing an idea that makes so much sense. Biomimicry has the earmarks of a successful meme, that is, an idea that will spread like an adaptive gene throughout our culture. Part of writing this book was my desire to see that meme spread and become the context for our searching in the new millennium.

I see the signs of nature-based innovation everywhere I go now. From Velcro (based on the grappling hooks of seeds) to holistic medicine, people are trusting the inscrutable wisdom of natural solutions. And yet I wonder, why now? Why hasn't our culture always rushed to emulate what obviously works? Why are we becoming nature's protégés at this late date?

THE STORM BEFORE THE CALM

Though it seems perfectly sensible to echo our biological ancestors, we have been traveling in just the opposite direction, driven to gain our independence. Our journey began ten thousand years ago with the Agricultural Revolution, when we broke free from the vicissitudes of hunting and gathering and learned to stock our own pantries. It accelerated with the Scientific Revolution, when we learned, in Francis Bacon's words, to "torture nature for her secrets." Finally, when the afterburners of the Industrial Revolution kicked in, machines replaced muscles and we learned to rock the world.

But these revolutions were only a warm-up for our real break from Earthly orbit—the Petrochemical and Genetic Engineering Revolutions. Now that we can synthesize what we need and rearrange the genetic alphabet to our liking, we have gained what we think of as autonomy. Strapped to our juggernaut of technology, we fancy ourselves as gods, very far from home indeed.

In reality, we haven't escaped the gravity of life at all. We are still beholden to ecological laws, the same as any other life-form. The most irrevocable of these laws says that a species cannot occupy a niche that appropriates all resources—there has to be some sharing. Any species that ignores this law winds up destroying its community to support its own expansion. Tragically, this has been our path. We began as a small population in a very large world and have expanded in number and territory until we are bursting the seams of that world. There are too many of us, and our habits are unsustainable.

But I believe, as many have before me, that this is just the storm before the calm. The new sciences of chaos and complexity tell us that a system that is far from stable is a system ripe for change. Evolution itself is believed to have occurred in fits and starts, plateauing for millions of years and then leaping to a whole new level of creativity after crisis.

Reaching our limits, then, if we choose to admit them to ourselves, may be an opportunity for us to leap to a new phase of coping, in which we adapt to the Earth rather than the other way around. The changes we make now, no matter how incremental they seem, may be the nucleus for this new reality. When we emerge from the fog, my hope is that we'll have turned this juggernaut around, and instead of fleeing the Earth, we'll be homeward bound, letting nature lead us to our landing, as the orchid leads the bee.

IN VIVO GENIUS

It may be a troubled conscience that is pushing us toward home, say the biomimics, but the critical mass of new information in the natural sciences is providing an equally important pull. Our fragmentary knowledge of biology is doubling every five years, growing like a pointillist painting to a recognizable whole. Equally unprecedented is the intensity of our gaze: new scopes and satellites allow us to witness nature's patterns from the intercellular to the interstellar. We can probe a buttercup with the eyes of a mite, ride the electron shuttle of photosynthesis, feel the shiver of a neuron in thought, or watch in color as a star is born. We can see, more clearly than ever before, how nature works her miracles.

When we stare this deeply into nature's eyes, it takes our breath away, and in a good way, it bursts our bubble. We realize that all our inventions have already appeared in nature in a more elegant form and at a lot less cost to the planet. Our most clever architectural struts and beams are already featured in lily pads and bamboo stems. Our central heating and air-conditioning are bested by the termite tower's steady 86 degrees F. Our most stealthy radar is hard of hearing compared to the bat's multifrequency transmission. And our new "smart materials" can't hold a candle to the dolphin's skin or the butterfly's proboscis. Even the wheel, which we always took to be a uniquely human creation, has been found in the tiny rotary motor that propels the flagellum of the world's most ancient bacteria.

Humbling also are the hordes of organisms casually performing feats we can only dream about. Bioluminescent algae splash chemicals together to light their body lanterns. Arctic fish and frogs freeze solid and then spring to life, having protected their organs from ice damage. Black bears hibernate all winter without poisoning themselves on their urea, while their polar cousins stay active, with a coat of transparent hollow hairs covering their skins like the panes of a greenhouse. Chameleons and cuttlefish hide without moving, changing the pattern of their skin to instantly blend with their surroundings. Bees, turtles, and birds navigate without maps, while whales and penguins dive without scuba gear. How do they do it? How do dragonflies outmaneuver our best helicopters? How do humming-birds cross the Gulf of Mexico on less than one tenth of an ounce of fuel? How do ants carry the equivalent of hundreds of pounds in a dead heat through the jungle?

These individual achievements pale, however, when we consider the intricate interliving that characterizes whole systems, communities like tidal marshes or saguaro forests. In ensemble, living things maintain a dynamic stability, like dancers in an arabesque, continually juggling resources without waste. After decades of faithful study, ecologists have begun to fathom hidden likenesses among many interwoven systems. From their notebooks, we can begin to divine a canon of nature's laws, strategies, and principles that resonates in every chapter of this book:

> Nature runs on sunlight.
> Nature uses only the energy it needs.
> Nature fits form to function.
> Nature recycles everything.
> Nature rewards cooperation.
> Nature banks on diversity.
> Nature demands local expertise.
> Nature curbs excesses from within.
> Nature taps the power of limits.

A CAUTIONARY TALE

This last lesson, "tapping the power of limits," is perhaps most opaque to us because we humans regard limits as a universal dare, something to be overcome so we can continue our expansion. Other Earthlings take their limits more seriously, knowing they must function within a tight range of life-friendly temperatures, harvest within the carrying capacity of the land, and maintain an energy balance that cannot be borrowed against. Within these lines, life unfurls her colors with virtuosity, using limits as a source of power, a focusing mechanism. Because nature spins her spell in such a small space, her creations read like a poem that says only what it means.

Studying these poems day in and day out, biomimics develop a high degree of awe, bordering on reverence. Now that they see what nature is truly capable of, nature-inspired innovations seem like a hand up out of the abyss. As we reach up to them, however, I can't help but wonder how we will use these new designs and processes. What will make the Biomimicry Revolution any different from the Industrial Revolution? Who's to say we won't simply steal nature's thunder and use it in the ongoing campaign against life?

This is not an idle worry. The last really famous biomimetic invention was the airplane (the Wright brothers watched vultures to learn the nuances of drag and lift). We flew like a bird for the first time in 1903, and by 1914, we were dropping bombs from the sky.

Perhaps in the end, it will not be a change in technology that will bring us to the biomimetic future, but a change of heart, a humbling that allows us to be attentive to nature's lessons. As author Bill McKibben has pointed out, our tools are always deployed in the service of some philosophy or ideology. If we are to use our tools in the service of fitting in on Earth, our basic relationship to nature— even the story we tell ourselves about who we are in the universe— has to change.

The ideology that allowed us to expand beyond our limits was that the world was put here exclusively for our use. We were, after all, the apex of evolution, the *pièce de résistance* in the pyramid of life. Mark Twain was amused by this notion. In his marvelous *Letters to the Earth*, he says that claiming we are superior to the rest of creation is like saying that the Eiffel Tower was built so that the scrap of paint at the top would have somewhere to sit. It's absurd, but it's still the way we think.

Where I live in the mountains of western Montana, a huge controversy is brewing about whether grizzly bears should be reintroduced to the wilderness area that sprawls outside our door. It's an issue that makes people scoop up their kids and get out their guns. The anti-reintroduction folks say they don't want to have to "take precautions" when they go hiking or horsepacking, meaning they don't want to have to worry about becoming a meal for a grizzly. No longer top banana, they would have to accept being part of another animal's food chain, a life-form on a planet that might itself be a life-form.

The rub is, if we want to remain in Gaia's good graces, that's exactly how we have to think of ourselves, as one vote in a parliament of 30 million (maybe even 100 million), a species among species. Although we are different, and we have had a run of spectacular luck, we are not necessarily the best survivors over the long haul, nor are we immune to natural selection. As anthropologist Loren

Eisley observed, all of the ancient city-states have fallen, and while "the workers in stone and gold are long departed," the "bear alone stands upright, and leopards drink from the few puddles that remain." The real survivors are the Earth inhabitants that have lived millions of years *without consuming their ecological capital*, the base from which all abundance flows.

NOSTOS ERDA: RETURNING HOME TO EARTH

I believe that we face our current dilemma not because the answers don't exist, but because we simply haven't been looking in the right places. Moi, upon leaving Washington, D.C., where he had seen hot showers, *The Washington Post*, and televised baseball for the first time, said merely, "There is not very much to learn in the city. It is time to walk in the forest again."

It is time for us as a culture to walk in the forest again. Once we see nature as a mentor, our relationship with the living world changes. Gratitude tempers greed, and, as plant biologist Wes Jackson says, "the notion of resources becomes obscene." We realize that the only way to keep learning from nature is to safeguard naturalness, the wellspring of good ideas. At this point in history, as we contemplate the very real possibility of losing a quarter of all species in the next thirty years, biomimicry becomes more than just a new way of looking at nature. It becomes a race and a rescue.

It's nearly midnight, and the ball is dropping—a wrecking ball aimed at the Eiffel Tower of squirming, flapping, pirouetting life. But at heart this is a hopeful book. At the same time that ecological science is showing us the extent of our folly, it is also revealing the pattern of nature's wisdom reflected in all life. With the leadership of the biomimics you will meet in the chapters that follow, I am hoping that we will have the brains, the humility, and the spirituality that are needed to hold back that ball and take our seat at the front of nature's class.

This time, we come not to learn *about* nature so that we might circumvent or control her, but to learn *from* nature, so that we might fit in, at last and for good, on the Earth from which we sprang. We have a million questions. How should we grow our food? How should we make our materials? How should we power ourselves, heal ourselves, store what we learn? How should we conduct business in a way that honors the Earth? As we discover what nature already knows, we will remember how it feels to roar like a jaguar—to be a part of, not apart from, the genius that surrounds us.

Let the living lessons begin.

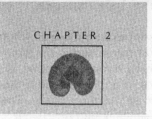

CHAPTER 2

HOW WILL WE FEED OURSELVES?

FARMING
TO FIT
THE LAND:
GROWING FOOD
LIKE A PRAIRIE

The native peoples who inhabited these lands long before us worshipped the Earth; they were educated by it. They didn't require schools and churches—their whole world was one.

—MICHAEL ABLEMAN, organic farmer, Goleta, California

How do we act on the fact that we are more ignorant than knowledgeable? Embrace the arrangements that have shaken down in the long evolutionary process and try to mimic them, ever mindful that human cleverness must remain subordinate to nature's wisdom.

—WES JACKSON, director of The Land Institute

I was at a friend's family reunion in Pipestone, Minnesota, a farming community in the squared-off, smoothed-out-straight corner of the state. Neat rows of wheat marched up to the doors of the Kingdom Hall, veered around the Quonset hut and its covey of pickups, then folded back together and marched on for miles.

Inside, we had hardly touched the Jell-O salad when news of the approaching weather snaked its way up and down the long banquet hall. Heads turned toward the southern doors and long-legged men began stepping over the benches that lined the tables. They bent down to whisper in the ears of other men, who excused themselves and swung their legs over and out. Through the doorframe, we could see a sky the color of carbon, a sky that would come off on your hands if you touched it.

I made my way out to the parking lot where men in their church clothes leaned against trucks dusted the same flat color as the soil. In silence they watched the weather come. A few lit cigarettes and winced as the clouds roiled, like smoke barreling before a runaway fire. "It's hail," said one of them finally. The others were already crushing butts and climbing into their Dodges and Chevys, peeling out to join the caravan. Wordlessly, the kids at my table collected silverware while their moms stacked plates and whisked tablecloths away. The festive air had turned funereal, and I had the feeling it wasn't the first time.

That storm turned into one of the hardest hails to hit southwestern Minnesota in a decade. What I realized then, viscerally, I knew already. Farmers are responsible for protecting their crops from things they cannot control. Today's farmer in southwestern Minnesota has a huge spread, and because the fields are planted in one species, one variety, and one growth stage, the losses, when they come, are catastrophic. Having put their eggs in one basket, they are at nature's mercy, caught in the crosshairs of drought, floods, pests, hail, and eroding soils. If anyone knows about being booted from the Garden of Eden, it's farmers.

What's amazing to observe is a natural grassland—a prairie— under the same kind of assault. Some of the grasses suffer, but most survive quite well, thanks to a perennial root system that ensures next year's resurrection. There's a hardiness about the plants in a wild setting. When you look at a prairie, you don't see complete losses from anything—you don't see net soil erosion or devastating pest epidemics. You don't see the need for fertilizers or pesticides. You see a system that runs on sun and rain, year after year, with no one to cultivate the soil or plant the seeds. It drinks in no excess inputs and excretes no damaging wastes. It recycles all its nutrients, it conserves water, it produces abundantly, and because it's chock-full of genetic information and local know-how, it adapts.

What if we were to remake agriculture using crops that had that same kind of self-sufficiency, that ability to live amiably with their fieldmates, stay in sync with their surroundings, build soil beneath them, and handle pests with aplomb? What would agriculture look like?

Well, that depends on where you live. Wes Jackson thinks it would look like a prairie. Jack Ewel thinks it would look like a tropical forest. Gary Paul Nabhan thinks it would look like a flood-washed desert. J. Russell Smith, were he alive today, would vote for a New England hardwood forest. The common theme is that the agriculture in an area would take its cue from the vegetation that grew there before settlement. Using human foods planted in the patterns of natural plant communities, agriculture would imitate as closely as possible the structure and function of a mature natural ecosystem. Threading our needle with the roots of such a stable system, we would sew up one of the deepest wounds on the planet— the gash made by till agriculture.

In many ways, this "farming in nature's image" movement is the most radical in this book, and perhaps the most important. As any economist would tell you, you can't eat widgets. Food is what is called a complementary; it's a given need that will always be with us, and despite what science fiction says about pill meals, there is no substitute.

Years after the hailstorm, I am once again in farm country, this time in Kansas, on my way to the country's premier enclave of agricultural researchers seeking to mimic nature's patterns. As I drive, a crew cut of wheat fields surrounds the car in all directions, as far as the eye can see. From the air it must look cut from a tool and die machine— straight rows of alternating green and brown, edged with an angularity foreign to living things. The soil beneath the stalks is plainly visible, all hint of weediness rounded up by chemical sprays. Nothing extraneous is allowed to grow here; everything has been stripped down to its least diverse form.

Whatever is left of the biotic community is harnessed and tuned to the production of one star: the cash crop. The fields have a factory-floor efficiency about them, and every now and then I see the floor managers, two stories up in Big Bud, Model 747 tractors, checking their six television monitors to see what's happening on the ground. Plumes of diesel smoke and soil billow behind their rigs, like live volcanoes spewing.

The soil plumes bring me back to a conversation I had at the Ravalli County Fair with a stoop-backed rancher who had farmed in Kansas during the Dust Bowl. He described windrows of soil so high that the cows used them like ramps to walk over the fences and out. "It was on account of plowing where we had no business plowing," he told me, "and what got lost on that wind, we never got back."

When I get lost in my wanderings around the wilderness areas of Montana, I usually don't realize it for a while. When I do, I have to rein in a skittery panic and try to think of how I got there, which landmarks I remember. Only then can I find the trail home. In agriculture, after being lost for the longest time, it's time to sit down and think.

HOW WE WOUND UP IN THE BOX CANYON OF INDUSTRIAL FARMING

It was ten thousand years ago that we split open the rich, ripe-smelling soil for the very first time. We saved a seed, planted it, and rejoiced when it grew up, spilling its harvest right into our hands. We celebrated our release from the gamble of hunting and gathering, and brought bumper crops of grain and babies into the world. The more babies we produced, the more land we had to put under production to feed our brood. We began to work the land harder and harder, moving up slopes and into other places we had "no business" farming. Although we improved our odds of a dependable larder, we had unwittingly stepped onto what plant breeder Wes Jackson calls a "treadmill of vigilance." The more we tamed and sheltered our crops, the more they depended on us for their survival.

By now, our crops are so far from the adaptive chutzpah of their wild ancestors that they can't do without us and our petrochemical transfusions of fertilizer and pesticides. In our quest for ever-increasing production, we removed their inborn defenses. We isolated them from mixed species groupings, narrowed their genetic diversity, and gutted the health of their soil.

Of these three, say historians of agriculture, eviscerating soil was our greatest misstep. Topsoil is essentially nonrenewable. Once eroded or poisoned, it can take thousands of years to rejuvenate itself. Rather than opt for a self-sufficient, perennial plant community that would batten down this black gold, we opted for the rip-roaring growth of annuals, which requires us to disturb the soil each year.

Each time we plow, we simplify the soil, taking away some of its capacity to grow crops. We break apart its intricate architecture and wreak havoc with the dream team of microfauna and microflora that glues it together into colloids, or clumps, of soil and organic matter. This clumping is vital; it leaves air channels like veins throughout the soil, giving water a way to sink down deep. Soils that are plowed too fine or packed too hard lose their colloids, and with them the art of retaining water. Parched air sucks the ground dry, and when the winds blow, talcum-powder topsoil coats the hoods of cars in town.

When rain strikes the hard pack, it can't shimmy down to the miles of thirsty roots as it should. Instead it glances off and runs in sheets, rills, and rivulets, murky and bloodstained, to the sea. The blood is soil, the living plasma of the Earth, sloughed off at a rate of five to one hundred tons per acre per year—a massive heist. Some Palouse Prairie wheat fields in Washington, on the shameful side of that equation, have the potential to lose one inch of topsoil every 1.6 years. In Iowa, up to six bushels of soil are washed out to sea for every bushel of corn produced.

What's left behind is a little deader as well as a little thinner. Behind the rest stop on Highway 7, I trespass a ways into a Kansas wheat field and bring up a handful of the bladed, pulverized, chemically amended soil. It's not chocolate-pudding black like the soil under the first plowed prairies must have been. It's beige and it doesn't smell as dank or fecund as it should—it doesn't smell like death and life commingled. The fungi that once wrapped their threads around rootlets to extend their reach, the brotherhoods of beneficial soil organisms, the bacteria that spun airborne nitrogen into food—they're all down to a skeleton crew, a shadow of their former selves. With the links among them severed, there is less "bootstrapping," less of the power that comes from several species working in biotic conspiracy to lift up the whole community.

The wildly fertile "postage stamp" prairies still scattered throughout the Great Plains give fragmentary testament to what we once had. In his eloquent book *The Grassland*, Richard Manning describes these vestiges as "pedestals carved by the plow." From the crown of some of these pedestals, once level with the land, you now have to drop down three feet to reach plowed soil. Such is what we have lost.

In other places, the scalp of the Earth is so thin that our plows are already mixing it with subsoil, which doesn't have the organic history that topsoil has. The grand larceny of harvest removes even more organic matter from these fields. Even in places where the stubble is plowed back in before planting, the nutrients are often wasted, pried away by hard rains before any plants are even visible. Over the years, these heists and the mistimed feedings add up to decreased fertility, a slow sterilization of our nation's real goose with the golden eggs. "Over a mere century of tilling the prairie soils of North America," says ecologist Jon Piper in his book *Farming in*

Nature's Image, ''we have lost one third of their topsoil, and up to 50 percent of their original fertility.''

Part of our loss can be traced to our fetish for production, our eagerness to turn an organic, nature-based endeavor into a factory: the farm as machine. Author and Kentucky farmer Wendell Berry says Europeans came to this continent with vision but not with sight—we couldn't see the value of what was right before us. We set to work removing the land's native dress and imposing a pattern of our own making. Exotic plants instead of indigenous ones, annuals instead of perennials, monocultures instead of polycultures. This disruption of a natural pattern, says Wes Jackson, is the definition of hubris.

Rather than looking to the land and its native peoples for instructions (what grows here naturally and why?), we issued arbitrary orders, expecting our farmland to fulfill many agendas, some of which had nothing to do with feeding people. Wheat, for instance, was leveraged to help us win the First World War. The European continent was overoccupied with fighting, and in many places, crops were neither planted nor harvested. To fill that void, we boarded battalions of newly motorized tractors and plowed our home soil right up to the Rockies, uprooting massive amounts of virgin prairie in what would later be called the Great Plow-up.

This was the finale of a movement that had begun with the first sodbusters and their steel-laminated moldboard plows, the only tools strong enough to break the tangle of prairie roots, some as stout as a homesteader's arm. It was considered backbreaking but heroic work, at least by white settlers. A Sioux Indian watching a sodbuster turn prairie roots skyward was reported to have shaken his head and said, ''Wrong side up.'' Mistaking wisdom for backwardness, the settlers laughed as they retold the story, ignoring the warning shots that fired with each popping root.

Having broken the prairie, we were ripe for the 1930s disaster of deep drought and relentless winds called the Dust Bowl. It got so bad our topsoil started showing up on the decks of ships a hundred miles off the Atlantic coast. One day in 1935, as officials in Washington, D.C., were hemming and hawing about what to do, a cloud of Great Plains soil fortuitously blew into town. A frightened Congress coughed, teared, and eventually created the Soil Conservation Service (SCS), an agency that would cajole and even pay farmers to conserve their soil. SCS agents were evangelical, and farmers were ready to repent, and together they were successful in getting our most erodible lands replanted to perennial, soil-holding grasses.

The institutional memory proved short, however, and when another world war had come and gone, we looked around and wondered why we weren't "using" every inch of the breadbasket. Earl Butz, the secretary of agriculture under Richard Nixon, reflected the nation's hubris by admonishing farmers to plow "fencerow to fencerow." Forgetting the lessons of the Dust Bowl, farmers filled in draws and bulldozed windbreaks, spending millions of federal dollars to obliterate what the SCS had spent millions of dollars planting.

We now had acres of new canvas on which to paint the next face of industrialized farming: the Green Revolution. In what was heralded as the answer to world starvation, breeders unveiled new hybrid strains of crops that promised phenomenal yields. Because of their hybrid nature, however, these new plants couldn't pass their genetic traits on to the next generation. So farmers around the world abandoned the time-honored (and ecologically prudent) tradition of seed saving and added a new expense to their ledgers: purchasing hybrid seeds.

The homogenization of fields spread rapidly. Varieties of crops that had once been used because they did well on a south-facing slope or were able to prosper in the Banana Belt or the Little Arctic regions of a state were forgotten. In places like India, where there were once thirty thousand land-tailored varieties of rice, their replacement by one super variety swept away botanical knowledge and centuries of breeding in one fell swoop.

Too late, farmers realized that touted yields were only promised, not guaranteed. In your part of the world, the fine print read, you may have to do a little goosing to get advertised yields—more water, more thorough tilling, more pest protection, more artificial fertilizer. But once the farmer next door had taken the bait and started to grow high-yielding varieties, you had to as well, so as not to be left behind. Together, like a slow pour over a large falls, we switched to a system of farming that mimicked industry, not nature.

Chasing economies of scale, experts advised farmers to get big or get out. Mechanization allowed them to "service" larger fields with less labor, but it meant steep capital investments: more land, bigger equipment, enormous debt. For the small operator, there was suddenly no room to dance in the margin, or to take care of your land the way you'd like. When you are in debt for a $100,000 combine, you can't afford to switch to alfalfa one year to rest the land. To hold the debt at bay, and to qualify for government subsidies, you have to farm volume.

We quickly went from growing food to sustain ourselves to growing so much food it became a surplus—an export item and a political tool. The farm became just another factory producing another product that would keep the United States in the global catbird seat. The internal controllers, those farmers with their ears to the land, determined to pass on good fertile soil to their progeny, gave way to remote-distance controllers—agribusiness and public policy.

To serve these "distance princes," as *Grassland* author Richard Manning puts it, industrial farmers abandoned traditional ways of managing their lands, such as rotating crops, liming and fertilizing with animal manure, or producing a diversity of products in case one crop failed. Instead, they "focused" their farms—selling off their livestock and switching to one species grown in continuous cropping, which is, in effect, continuous robbing. They propped up flagging soil fertility with artificial nitrogen fertilizer produced with natural gas. Weed competition was quelled with herbicides, another petroleum product, while oil-based chemicals were used as a prophylactic against pest outbreaks (which by now were extreme, thanks to acres of identical plants with identical vulnerabilities). Suddenly, for the first time in ten thousand years of agriculture, farmers were beholden to the protection ring of petroleum and chemical companies, and were said to be growing their crops not so much in soil as in oil.

Once on that treadmill, the feedback loops began. Weeds and pests are wily by nature, and even if you spray them one year, not all of them will die. Those that manage to hack an immunity explode the next year, requiring even heavier doses of biocides. In the escalating war of "crops and robbers," the more you spray, the more you have to spray.

Who's winning? Since 1945, pesticide use has risen 3,300 percent, but overall crop loss to pests has not gone down. In fact, despite our pounding the United States with 2.2 billion pounds of pesticides annually, crop losses have *increased* 20 percent. In the meantime, more than five hundred pests have developed resistance to our most powerful chemicals. On top of that bad news, the last thing we want to hear is that our soils are also becoming less productive. Our answer has been to rocket-boost fertility with 20 million tons of anhydrous ammonium fertilizer a year—as many as 160 pounds per person in this country alone.

Recently, the protection racket has jumped to a whole new level of menace. Tune in to TV in an agricultural state, and you'll see a slick commercial for a crop seed that comes pretreated with a herbicide that kills weeds but doesn't harm the growing seedling. Because the plant has been specially bred to grow unscathed by that brand of herbicide and none other, the company is assured future sales. There's something unsavory about this. A dependence forms, and product loyalty is instilled with no question about the wisdom of using that product. Evidently, this latest move has been in the offing for quite some time. According to a December 1982 *Mother Jones* article by Mark Schapiro, at least sixty U.S. seed companies were sold between 1972 and 1982, all of them to chemical and petroleum companies. At last count, sixty-eight companies have plans to introduce their own seed/herbicide combos. Good news, they say: Now that farmers don't have to worry about seedlings suffering from year-to-year herbicide carryover (which used to limit herbicide use), they can use as much as they want.

This is the kind of news that should worry all of us. At last count, leaching pesticide residues made agriculture the number-one polluting industry in this country. At stake is groundwater, which supplies half the U.S. population with its drinking supply, and which is nearly impossible to clean once contaminated. Farm families already know about contamination. Recent studies have shown that people living in rural parts of Iowa, Nebraska, and Illinois are likely to have pesticide residues in their wells, and to have higher than normal risks of developing leukemia, lymphoma, and other cancers. Nitrate levels (from fertilizer) in the drinking water of many farm communities also exceed federal standards, which may be why miscarriage rates in farm families are unusually high.

Nitrates are not the only thing draining from farmland. Money is, too. In 1900, if you put a dollar's worth of material and energy inputs into your farm, you'd produce $4.00 worth of crops, an input-to-production ratio of 1:4. Today, even though we produce more food, our genetically pauperized, oil-hungry crops cost more to grow. It takes $2.70 worth of oil-based inputs to produce $4.00 worth of crops, an input-to-production ratio of only 1:1.5.

Moreover, because of the crops and robbers feedback effect, we will continue to need more and more inputs. Already, Cornell University ecologist David Pimentel reckons that society spends ten kilocalories of hydrocarbons to produce one kilocalorie of food. That means each of us eats the equivalent of thirteen barrels of oil a year.

Author Richard Manning cuts through these statistics to ask the important question: When you have a system that is one part farmer and nine parts oil, who do you think will have the ultimate power? Not small farmers, and certainly not the landscape.

According to data collected by Iowa State University in 1993, most farm families now rely on off-farm revenues for one half of their income. Those who don't make it wind up selling to those with ready cash—corporations, syndicates, investors. This spiral leads to fewer family farms and a brain drain from the rural countryside, a tragedy that Wes Jackson calls "fewer eyes per acre." Already, 85 percent of our food and fiber comes from 15 percent of our farms. These megafarms are hardly what Thomas Jefferson envisioned when he saw a nation of yeoman farmers tending their 160 acres, beholden to no one.

What's most dangerous about this dependency—the crops on us and us on petroleum—is that it keeps us too busy to think what the real problems might be. Fertilizer, for instance, masks the real problem of soil erosion caused by a till agriculture of annuals. Pesticides mask a second real problem: the inherent brittleness of genetically identical monocultures. Money borrowed to pay for the fossil-fuel inputs masks a third real problem: the fact that industrial agriculture not only destroys the soil and water, it strangles rural communities. Though we don't want to admit it, our farms have become factories owned by absentee interests. With our help, they are liquidating the ecological capital that took the prairie five thousand years to accumulate. Every day, our soil, our crops, and our people grow a little more vulnerable.

What I want to know is, how long can our denial hold?

Before I get too deep into despair, I remind myself that I am headed to meet the one group of researchers who have stepped from denial's shadow and made it their business to expose the crumbling foundations of this system. The people at The Land Institute—fifteen staff members, nine interns, and three staff volunteers—are committed to devising an agriculture that is, in Director Wes Jackson's words, "more resilient to human folly." On one of my driving breaks, I reread The Land Institute's literature, and its quiet, determined tone assures me as much as it amazes me. At core these researchers are farmers, and they think there is nothing more sacred than the pact between humans and the land that gives them their food. But they are also realists, and it's made them revolutionaries. They're not afraid to acknowledge that it's not just a few problems *in* agriculture that need overhauling. It's the problem *of* agriculture itself.

The problem of agriculture is an old and pervasive one, explains Wes Jackson in a series of books including *New Roots for Agriculture, Altars of Unhewn Stone,* and *Meeting the Expectations of the Land.* It comes from an insistence on decoupling ourselves from nature, from replacing natural systems with totally alien systems, and from waging war on, rather than allying ourselves with, natural processes. The result has been a steady loss of ecological capital—the erosion and salting of soil, the steady domesticating and weakening of our crops. To find our way back, says Jackson, we have to remember what the ancestors of "our" crops were like in their own element.

Once wild creatures, our agricultural charges were shaped by an ecological context that bears little resemblance to our farming. Their natural ecosystems ran on sunlight, sponsored their own fertility, fought their own pest battles, and held down, even *built,* soil. But long ago, plants were removed from the original relationships they had with their ecosystems and pressed into our service. Now, writes Jackson, "Our interdependency has become so complete that, if proprietorship is the subject, we must acknowledge that in some respects they own us." To break this codependent cycle, we have to stop fighting our crops' battles and instead raise hardy crops in a farming system that brings out their natural strengths.

THE PARABLE OF THE PRAIRIE

"Essentially, we have to farm the way nature farms." Wes Jackson, sixty-year-old fourth-generation Kansas farmer and modern-day Cain-raiser, arrived at that simple conclusion years ago, before he had the language to speak of it. It was his sixteenth summer and he was away from his family's Kansas farm, roping and riding on his cousin's cattle ranch in South Dakota. He was amazed that no one planted or tended it, yet the grass came up year after year, drought or no drought, through snow and blistering sun. There were rattlesnakes coiled right in the middle of it, and burrowing owls standing sentry outside their holes. "There was a rightness to it all," he says now.

Another good rain fell while Jackson was working toward his Ph.D. in genetics at North Carolina State. His adviser, Ben Smith, popped his head in the door one evening and declared: "We need wilderness as a standard against which to judge our agricultural practices." With this, the seed coat split, and a slow root began to burrow.

When Jackson was thirty-seven, on the fast track to tenure after writing a successful text called *Man and the Environment*, he got uneasy. Though he had an enviable post as creator of the Environmental Studies Department at California State University in Sacramento, he felt he wasn't where he was meant to be. To the astonishment of his colleagues, he and his wife, Dana, packed up their three kids and returned home to Kansas. They moved into a partially earth-sheltered house that they had built along the Smoky Hill River, and in 1976, they began a school that focused on sustainable living practices. That school would become The Land Institute, a nonprofit research organization devoted to "an agriculture that will save soil from being lost or poisoned while promoting a community of life at once prosperous and enduring." This new agriculture would take wilderness as its model, nature as its measure.

In Kansas, the wilderness was tallgrass prairie, the natural expression of the underlying layers of soil, the carnival of weather, the licking of fire, and the grazing of elk and bison. Prairie is what Kansas land wants to be, but for the most part, is no more.

I am startled, then, by what I see when I turn down Water Well Road to The Land Institute. With no warning, the bristle of wheat fields yields to a softer ensemble of wild-haired plants, stems akimbo, saturated with color and raucous with flowers and tasseled stalks. As I watch, wind enters like a dancer onto a crowded floor, parting the crowd, causing a bobbing and dodging of plants in its wake. The whole thing sways crazily for a moment, then settles in a perfect hush, like a band ending a jam by feel.

A sign by the road says that this is The Wauhob, a prairie miraculously spared a sodbusting, probably because it was up gradient, and hard to get plows to. My car literally rolls to a stop as I gawk, so welcome is this sight after the acre upon acre of ramrod efficiency I've driven through. From where I am now, I can see both wheat field and prairie, and it's like a visual parable—Jacob and Esau, cut from the same cloth but of very different character. One is the expression of imposed will, the other the expression of the land's will. An understanding intern spots me and interrupts his organic gardening chores to give me directions to the office.

The Land Institute headquarters is a modern brick house that was once home to an older couple. The bedrooms are now offices, and there's a kitchen and fireplace in the meeting room, where a dozen local women are stuffing envelopes and drinking coffee when I arrive. In "The Land's" twenty years of existence, its original 28 acres have grown to 270, an amazing accretion considering that this nonprofit operates solely on private funding and has never gone into debt.

Ecologist Jon Piper greets me in the foyer and asks about my drive, all the while migrating toward the door, as eager as I am to get out into the prairie. Piper is in his late thirties, bespectacled and bearded, with a quiet forbearance for visitors like me. He knows that what I experience here, my dip in the prairie sea, will be as important as what we say to one another. "We'll make a conceptual loop," he tells me. "Starting where all our thinking starts."

As we wade into the knee-high Wauhob, Piper comes to life, unconsciously bending and turning the heads of the plants as he talks, like a teacher touching the heads of students as they work. Though never planted by human hands, the prairie is choked with blossoms, grasses gently pouring over, seeds setting, new shoots growing, runners crisscrossing the earth in a web of decay, growth, and new life. There is no hint of hail damage or drought wilt, no such thing as weeds. Every plant—231 species in this patch alone—has a role and cooperates with linked arms with the plants nearby. I see diversity of form—grasses splaying upward to different heights and widths, a sunflower's bold expanse, a legume's dark leaflets, fernlike in their repetition.

Piper talks about the plants as if they are neighbors in a community—the nitrogen fixers, the deep-rooted ones that dig for water, the shallow-rooted ones that make the most of a gentle rain, the ones that grow quickly in the spring to shade out weeds, the ones that resist pests or harbor heroes such as beneficial insects. He also points out the butterflies and bees, the pollinators with wagging tongues, spreading rumors from one plant to another.

Beneath this unruly mob lies 70 percent of the living weight of the prairie—a thick weave of roots, rootlets, and runners that captures water and pumps nutrients up from the depths. A single big bluestem will have twenty-five miles of this fibrous plumbing, eight miles of which will die and be reborn each year. These root remains, together with the leaves shed from above, will fall into the welcoming jaws of a miniature zoo—ants, springtails, centipedes, sowbugs, worms, bacteria, and molds. There are thousands of species in a single teaspoon, all tunneling, eating, and excreting, conditioning the soil crumb by crumb. Through their magic, dissolved nutrients are released to thirsty roots or stored in humus—the tilth that transforms the prairie into a living sponge.

The character of this belowground world is an expression of the bedrock, organic matter, rainfall, temperature, light conditions, and most important, the plant and animal community above. Pluck or plant something new and you change the microecology slightly. Plow, spray, and harvest every year, and you change it plenty. Some of the organisms you lose might be those that sponsor fertility, or help stave off insect and disease attacks, or produce hormones that tell a flower to unfurl or a root to push its snout deeper into the soil. It takes years to tune such an orchestra of microhelpers, but just moments to silence it.

The secret of the prairie is its ability to maintain both aboveground and belowground assemblies in a dynamic steady state. It's not the fact that nothing changes on the prairie (patches are always pulsing with change), but that the changes are never catastrophic. A prairie keeps pest populations in check, rebounds gracefully from disturbance, and resists becoming what it is not—a forest or a weed garden.

"Our goal at The Land Institute is to design a domestic plant community that behaves like a prairie, but that is predictable enough in terms of seed yield to be feasible for agriculture," says Piper. To illustrate, he heads downslope from The Wauhob to stand in the zone between the prairie and wheat field I saw earlier. "Down there is our current agricultural ideal; we know it isn't sustainable, mainly because it loses soil and requires nonrenewable inputs. Up where you are, we have a sustainable ideal, but it won't feed us. Conceptually, we'd like to be somewhere in here, between the controlled rigidity of the wheat field and the wildness of the prairie."

It's a concept that I'd read about in chaos and complexity literature. There exists a sweet spot between chaos and order, gas and crystal, wild and tame. In that spot lies the powerfully creative force of self-organization, which complexity researcher Stuart Kaufmann calls "order for free." Tropical agroecologist Jack Ewel also alludes to this free ordering when he says, "Imitate the vegetative structure of an ecosystem, and you will be granted function."

As the first step toward an agriculture that organizes itself into arrangements of strength, Piper's job was to ascertain just what it was about the prairie's structure that made it so tough. Is there a rule of thumb about which categories of plants consistently show up on a prairie roster, and what ratios they are in? Does it matter where they grow in relation to one another? In search of answers, Piper read everything he could about prairie ecology, and then spent seven glorious summers up to his eyebrows in wild pastures. He and his interns actually took scissors and clipped and bagged all the vegetation in certain plots. They identified each and every plant, separated them out into piles, and then dried and weighed them to find out what grew there. Through wet years and dry years, in rich soil and poor, Piper found that prairies do have a pattern that repeats itself, an order in the seeming chaos.

"The first thing that strikes us," says Piper, "is that ninety-nine point nine percent of the plants are perennials. They cover the ground throughout the year, holding the soil against wind and breaking the force of raindrops. Hard rain hits this canopy of plants and it either runs gently down the stems or it turns into a mist. By contrast, when rain hits row crops, it strikes exposed soil, packs it, then runs off, taking precious topsoil with it." Researchers have actually measured the difference; in identical downpours, they found that you get eight times as much runoff from a wheat field as from a prairie.

"Prairies just soak up a big rain," says Piper. "I can come out here hours later, and The Wauhob still squishes when I walk on it."

Besides being great sponges, perennials are also self-fertilizing and self-weeding. Thirty percent of their roots die and decay each year, adding organic matter to the soil. The remaining two thirds of the roots overwinter, allowing perennials to pop open their umbrella of vegetation first thing in the spring, long before weeds can struggle up from seed. As we walk through a particularly dense patch of prairie, Piper crows, "See? You wouldn't have a chance in there if you were a weed.

"The second thing that strikes us about the prairie is its diversity," Piper says. "We have two hundred and thirty-odd species right here on this knob—not just one species of warm-season grass, but forty species. Not just one nitrogen-fixing legume, but twenty or thirty. That means that there will always be some species or some variety of a species that can do well in our highly variable Great Plains climate. I've been out here in dry years when the grasses barely reach your knee and there's yucca everywhere. Other years, after plenty of rain, you and I could stand three feet apart and not be able to see each other through the big bluestem. The species composition remains the same, but different species excel in different years."

Diversity is also the cheapest and best form of pest control. "Many pests tend to specialize on one host plant species, so when there's a diverse mix, pests have a harder time finding their target plant. Even if they manage to touch down somewhere in the field, the attack troops don't get very far. Disease spores may blow onto the wrong plant, or insect young may crawl into the wrong bud. With a diverse offering, attacks die down before they become epidemics."

The third signature of the prairie is its four classic plant types: warm-season grasses, cool-season grasses, legumes, and composites. Cool-season grasses come up early, set seed, and then bow out of the way, allowing warm-season grasses such as big bluestem to rule the rest of the season. Legumes such as cat's-claw, sensitive brier, and leadplant fix their own nitrogen, fertilizing the prairie with their bodies. Composites, such as goldenrod, asters, and compass plants, can flower anytime throughout the season. Although these four "suits" may vary in proportion from place to place, Piper found them in every prairie he waded through.

"Learning the secrets of the prairie gave us a goal to shoot for as we sifted through the countless combinations of plants that would qualify as prairie mimics in our agriculture. We knew we needed perennial grains grown in a polyculture, with the four suits of the prairie represented. The only question was how *many* different species in each group will we have to plant? Since it's impractical to have an agriculture with two hundred species, how much diversity will we need to get functional stability? Our intuition told us that we would probably have to plant many more species than we need and let the assemblage shake down over a few years to the handful that would provide human food. Just about then, 'community assembly' studies started to show up in the literature, and they suggested that you could get persistent communities containing as few as eight species. That was encouraging to us."

Breeding eight perennial crop species from scratch looks more feasible than breeding two hundred, but it's still a daunting challenge. Today, most of the food eaten around the world comes from only about twenty species, and none of them are perennials! Some began as perennials, but over the ten-thousand-year odyssey of plant breeding, we systematically removed their hardy perennial traits, marching right by the sweet spot between wild and tame, and domesticating them until they were annual by nature.

A story is told about the moment Wes Jackson realized the full extent of this unhappy extreme in agriculture. Shortly after starting his school, Jackson took his students on a field trip to the eight-thousand-acre Konsa Prairie near Manhattan, Kansas. One of them asked the innocent question, "Are there any perennial grains?," and it made Jackson think. When he got back, he drew up a list of all the crops he could think of, separating them into either annual or perennial, herbaceous or woody, vegetative or seed/fruit yielding. To his surprise, there were crops that fit into almost all the categories, but there was a glaring blank in the space for HERBACEOUS, SEED-YIELDING PERENNIAL. It was a revelation in black-and-white.

PERENNIAL OPTIMISTS

Jackson and his staff started tearing apart the literature—surely someone must have done some plant breeding on perennial grains. They were disturbed to find that no one, save some folks looking at animal forage, had studied seed-yielding perennial grasses or legumes or composites. The reason?

"It was a nonstarter for career-oriented scientists," says Jackson. "The common wisdom was that perennials, which spend most of their energy belowground, could never be made to produce copious seeds [the part that humans eat]. If they were to yield more seeds, the thinking went, there would be a trade-off belowground, and they'd lose their ability to overwinter."

Jackson, who'd made a career of bucking conventional thought, said not so fast. The first question The Land Institute assigned itself was the one everyone else had skipped:

Can a perennial produce as much seed as an annual crop?

After two more years of library safaris and actual planting experience, The Land Institute staff was convinced that perennials *could* be bred to yield plentiful seeds without losing their perennial traits. Illinois bundleflower and wild senna, for example, were two wild perennials that, with absolutely no breeding, already approached the benchmark yield (the floor range) for wheat in Kansas: eight hundred pounds per acre. Considering that the wild relatives of some of our crops have undergone four-, five-, even twentyfold seed-yield increases at the hands of talented breeders, the chances of upping yields for these new crops were good.

The trick this time around would be to increase seed yield *without* stripping the plant of its wild hardiness. Curious to see what artificially increased seed yield would do to plant vigor, Jackson's daughter, Laura Jackson, a researcher at the University of Northern Iowa, conducted an experiment that showed that a plant need not sacrifice photosynthate—the ability to feed itself—when it puts out lots of seeds. In short, the trade-offs were not as strict as everyone imagined, and it seemed that the chimera The Land Institute wanted to create was well within the realm of the possible.

In 1978, the staff embarked on the painstaking process of breeding crops for the domestic prairie. They would have to possess not only hardiness but also "crop character"—qualities like good taste and ease of threshing. Since the breeding of most of the crops we eat today was fairly well wrapped up by Abraham's time, crop domestication of this sort was a brave new venture. The precedent for this work completely disappears when you consider that Jackson and crew were shooting for crops that were dependable, but *not* dependent on us.

There were two ways they could wind up with a perennial grain—one, they could start with a wild perennial and boost its seed yield and crop character, or two, they could start with an annual that already had good crop character and cross it with a perennial wild relative to refresh its memory about how to survive the winter. Now all they needed were candidates.

Going on catalog descriptions of native perennials in each of the groups, they ordered nearly five thousand different types of seed from governmental seed collections and planted them in the undulating fields by the Smoky Hill River. Those that survived well in Kansas weather and had a whiff of a hope for high seed yield became candidates in their breeding program. They planted the seeds and waited anxiously, as farmers do, to see how the plants matured. Besides seed yield, they were also looking for agronomic characteristics important to a farmer: reduced seed shattering (so seed heads don't break open and spill their grain before harvest), uniform time of maturity, ease of threshing, and large seed size.

The four most promising candidates for perennial domestication turned out to be eastern gamagrass (*Tripsacum dactyloides*), a sprawling warm-season grass that is a relative of corn; Illinois bundleflower (*Desmanthus illinoensis*), a legume that grows tall and produces a baby rattle of seed pods; mammoth wildrye (*Leymus racemosus*), a stout cool-season relative of wheat that the Mongols used to feast on when drought claimed their annuals; and Maximilian sunflower (*He-*

lianthus maximilianii), a composite that yields oil-rich seeds, which could be pressed to create vegetable oil diesel fuel for tractors. The second approach—starting with an annual and hybridizing it with a perennial—led to the mix of milo grain sorghum, which is already used as a crop, and perennial Johnsongrass.

Now that The Land has its lineup, the breeding has begun in earnest. The very best individuals from each species are grown together in one plot so that they can cross-pollinate. When two promising strains "mate," the hope is that even more bodacious offspring will follow. The seeds from each trial are planted out (in various kinds of soil to make sure the differences are truly genetic, or inheritable, and not just environmental), and the best individuals are selected to cross-pollinate once again. This process is repeated until the improvements due to crossing show signs of diminishing returns. Only then will the breeders call them good and begin the fine-tuning process to bring out each strain's best features.

So far, optimism at The Land is high, which means a slightly deeper nod from the incredibly modest Jon Piper when I ask whether he's pleased with their progress. He walks me among the monoculture and polyculture plots where the best of the best are showing their stuff. Some collections of eastern gamagrass are bravely resisting various leaf diseases, and certain collections of bundleflower and gamagrass are yielding well despite some drought. The most vigorous crosses between Johnsongrass and grain sorghum are showing both high seed yield and good rhizome production. (Rhizomes are the underground runners that allow plants to store starch for winter, and thereby survive.)

In terms of seed yield, there are already some superstars. Even though its food value has yet to be explored, says Piper, Illinois bundleflower is yielding seed quantities that approximate the typical yield of nonirrigated soybeans in Kansas. For eastern gamagrass, which can be ground into a cornmeal and baked into a palatable bread, the potential to improve seed yields is great, thanks to a variety that was discovered along a Kansas roadside. The collector noticed that instead of the normal flower stalk, which is composed of about one inch of female flowers topped by four inches of male flowers, this sport had all female parts (which turn into seeds) except at the very tip. If all yielded, the sport could produce up to four times the normal amount of seeds. As Piper shows me one of the stalks, I notice that the female organs are green. "Exactly," he says.

"That means they can photosynthesize and pay their own bills, meaning the plant won't necessarily have to trade off fewer roots in order to support more seeds. That's what we'll be trying to show."

By taking on this perennial-grain breeding challenge, folks at The Land Institute were already mucking in that part of the map that warned "serpents lie here." While they were at it, they thought they'd try for another first by choosing most of their candidates from native stock. (The only plant in their lineup that isn't native is mammoth wildrye.) Though native stock seems an obvious choice, it hasn't been to other breeders. Most of our crops are exotics, brought over in our traveling bundles from Mexico and Europe. The only native plants that we have ever domesticated in this country are sunflowers, cranberries, blueberries, pecans, Concord grapes, and Jerusalem artichokes. The Land Institute is trying to lengthen this short list, knowing that natives are tuned through evolution to sing in harmony with the melody of local conditions.

While coaxing agronomic manners from these plants will be a Pygmalion task, growing them in monocultures at least gives breeders a chance to compare apples with apples. Unfortunately, says Jackson, we can't stay with monocultures. The real Holy Grail is to grow them in *polyculture*—mixed species plots—since, as nature has shown us, only polycultures are able to pay their own bills.

POLYCULTURE SHOCK

Polyculture is not music to a breeder's ears. When you are working in a polyculture, you take all the difficulties that you encounter in monoculture breeding and multiply them. You are not only selecting for high yields, large seed size, uniform maturation time, easily threshed seeds, low shattering, winter hardiness, disease and pest resistance, and climate tolerance, but also for compatibility—a plant's ability to perform well or even exceed performance when grown next to other plants.

The Land Institute staff was essentially faced with designing an agricultural dinner party, deciding who should be seated next to whom to maximize the beneficial interactions and minimize the detrimental ones. Nature arranges these kinds of matchups all the time through the slow culling of natural selection. Could The Land somehow mimic and speed up this process?

"The traditional scientific method offered one way to go about it," says Piper, "and we worked this way for a while—planting out seedlings in mixed plots, purposefully putting certain species next to others so we could investigate their interactions." The problem was that the number of possible combinations is astronomical, and not even a Mendelian monk's life would be long enough to try them all. Just as Piper and his colleagues started questioning this reductionist approach, they began to read about recent developments in the field of community assembly.

James Drake and Stuart Pimm of the University of Tennessee study what it takes to arrive at an assembly of species that remain in equilibrium, a condition farmers would obviously want for their domestic prairie. Unlike The Land staff, they do their experiments with ecosystems in a computer (artificial life) and with aquatic organisms in glass tanks (real life). They begin by adding species in various combinations and then letting them work out who will survive and in what ratio. Eventually, without intervention, the community shakes down into something that is both complex and persistent—order for free. "But we don't get order immediately," says Pimm. "We get it after a long period of adding species to communities and watching them come in, displace other species, and go extinct in their turn." In other words, having a history is what makes a community last.

In his famous "Humpty Dumpty" hypothesis, Pimm maintains that once you destroy a finished product of community assembly, such as a prairie, you can't just plant those same species and expect to put it back together again. There's no such thing as an instant prairie. A prairie restorationist must give the prairie a successional history, that is, actually grow the prairie over a trajectory of years. Some plants will blow in and others will drop out, but as those facilitating species change the soil and the fauna and flora around them, they *make it possible for the final assembly to be there*. They warm up the crowd for the real act.

"The question for us mortal scientists," says the understated Piper, "and for farmers who will someday grow diverse perennial grains, is how to get that order quickly. We're not in the business of creating prairies over a thousand years. What we want to do is build complex, persistent systems that shake down within a very few years."

Nor do they have a thousand years to do the research. What Piper and company have decided to try, in addition to their more reductionist experiments, is a "shakedown" like those that occur in

Pimm's and Drake's experiments. First, they laid out sixteen plots (sixteen meters by sixteen meters), then randomly broadcast seeds that represented the prairie's four "suits": warm-season grasses, cool-season grasses, legumes, and composites. In some plots they sowed only four species, in others eight, twelve, and sixteen. There are four replicates of each treatment. Half of the plots are being left alone to develop as they will, and the other half are called "replacement" plots. After two years, any species in the replacement plots that have dropped out or failed to germinate will be replaced. "We want to give our target species every opportunity to join the community," says Piper. "It may be that mammoth wildrye can't establish itself in the first year or the second year, but it can in the third."

They'll keep track of which species comes on line when, and which of the plots leads to their desired community first. All the while they will be tracking changes in the communities and looking for rules and patterns about how stable communities assemble. Within a few growing seasons, they want their target perennial grains to be well represented, and to yield abundantly year after year without weeding or seeding. If a few other noncrop species are present in the mix, so be it. "If a plant is consistently present, it probably plays a role in maintaining stability," says Piper. Eventually, the "recipe" or trajectory the researchers discover will be something they can offer to farmers.

Though they don't know all the particulars, Piper thinks a typical recipe may work like this: You throw in the recommended mix of species (more than you need), making sure that all important plant groups are represented. Then you sit back and watch the trajectory unfold. The trajectory might take five years, say, but you would be rewarded with a complex, persistent system.

"Right now, for instance, we're seeing a flush of annual weeds the first and second year. The fields look awful at first, like a total failure, but the perennial seeds are in there and by the second or third year, they just go *whoosh* and come into their own. Somehow, the environment filters out what works from what doesn't work, so you are left with the most stable combination. We're studying how this happens, and what steps we might take to help it to happen." As their plots mature, The Land Institute will be experimenting with various management techniques to favor perennial grains and to get the community to gel. The resultant recipe might include a recommendation to burn in year two, mow in year three, or graze livestock in year four. They'll also be thinking about the equipment that will be needed to harvest the different crops at different times of year.

"Farming with a crowd of perennials is going to be different," says Piper. "It'll be more like forestry in that you have to wait a while to get to a harvestable stage. Also as with forestry, you can't just start over each year. You can't decide to grow another crop because pests are bad or the weather doesn't cooperate. Instead, you'll have to plan up front for multiyear conditions—weather, markets, et cetera. Your best hedge against disaster is going to be variety, just as the prairie teaches—lots of paints in your palette so that no matter what the conditions, some species will still flourish."

Besides getting the domestic prairie to take, visionaries at The Land also want it to fulfill its promise agriculturally. It has to compete reasonably well with what farmers are now growing. The final three questions that occupy Piper and company have to do with the polyculture performance from that pragmatic point of view.

Can the polyculture yields stay even with or actually overyield those of monocultures?

Overyielding is the phenomenon by which a crop yields more per unit acre when it's growing in a polyculture than when it's in a monoculture. Turns out that plants grown next to different but complementary neighbors don't have to compete the way they do when grown next to an identical plant. They're not jostling root elbows for the water in a particular level, for instance. Nor are they competing for the same plane of sunshine. As a result, the members of a diverse community are actually capturing more resources (and yielding more) than they would under constant same-species competition.

The literature is replete with examples of overyielding when complementary annuals such as maize, beans, and squash are planted together. Piper's charge was to show that overyielding could happen with perennials as well. "Sure enough, we're seeing it," he says, letting a grin escape with this news. "Nineteen ninety-five was the fifth year of a study of polycultures of eastern gamagrass, wildrye, and Illinois bundleflower. When compared with their performances in monoculture, plants in mixtures have consistently overyielded."

Can the polyculture defend itself against insects, pests, and weeds?

Studies at The Land are showing that when plants are grown in bicultures and tricultures, they're better able to fight off insects and diseases than when they're grown in monocultures. It makes sense if you think about it. Plants defend themselves against insects with chemical "locks," and at most, an insect carries only one or two "keys" to the plants it is adapted to eat. An insect that finds itself in a field of nothing but its target plant is like a burglar with the key to every house in the neighborhood. In a polyculture, where all the locks are different, finding food is more of a chore. A mixed neighborhood is equally frustrating for diseases that specialize in one plant. A fungus may fester on an individual, but when it releases its spores, the leaves of invulnerable plants act as a flypaper, bringing the fungal rampage to a halt. That's why, although pests exist in prairie polycultures, you don't see the runaway decimation that you see in monocultures. Invasions are contained.

Just as with overyielding, most of the experimental evidence for resistance comes from studies on annual plants in polycultures. In 1983, Cornell biologists Steve Risch, Dave Andow, and Miguel Altieri reviewed 150 such studies and found that 53 percent of the insect pest species were less abundant in annual polycultures than in annual monocultures. Similarly, Australian ecologist Jeremy Burdon summarized 100 studies of two-component mixtures and found that there were always fewer diseased plants in the polyculture. So far, the same seems to hold true for the perennial polycultures planted at The Land. "In the third year of testing," says Piper, "we had a sudden buildup of beetles on bundleflower. But only in the monocultures. The bundleflower that was grown with gamagrass was fine. Polycultures also seem to reduce or delay the onset of maize dwarf mosaic virus, which can be a problem on eastern gamagrass." Farmers are especially intrigued by these results, since they seem to indicate that pesticides could be scaled back or even eliminated in polycultures. With the thought of pesticides gone, Piper and his colleagues began fantasizing about eliminating another petroleum-based crutch: nitrogen fertilizer.

Can the polyculture sponsor its own nitrogen fertility?

The question of how much nitrogen fertilizer a domestic prairie would need has not been definitively answered as of this writing. So far, though, signs are pointing to little or none. In experiments conducted with annuals, soil fertility always looks stronger in a polyculture, especially when legumes are in the plot. Tiny balls on the roots of a legume (such as Illinois bundleflower) are home to bacteria that have the ability to turn atmospheric nitrogen into plant food. As a result, legumes find a niche in nitrogen-poor soils, thriving where other plants falter. Plants growing near the self-sufficient legumes may also benefit from stored nitrates that return to the soil when the legume sheds a leaf, turns over a portion of its roots, or lays down its last.

In initial investigations of polycultures that include Illinois bundleflower, Piper found that, as predicted, bundleflower can grow beautifully and yield well even in poor soil, leaving the soil character actually improved. As Piper relates in scientific papers, "The soil nitrate concentration in four-year-old Illinois bundleflower stands at the poorer soil site was nearly identical to that on the better soil site despite very different initial nitrogen conditions." Growing legumes is like having a crop that yields a harvest *and* simultaneously fertilizes your field. Which is why, of course, no prairie would be without them.

Despite the promise of The Land Institute's work, we're a long day away from finding gamagrass bread in our local supermarkets— twenty-five to fifty years, if these researchers are the only ones working. "We're at the Kitty Hawk stage," says Jackson. "We've demonstrated the principles of drag and lift, but we're not yet ready to fly people across the Atlantic in a Boeing seven-forty-seven."

They are ready to make some thrilling claims, however. In Eugene, Oregon, I saw Wes Jackson give an audience goosebumps with this statement: "After seventeen years of scientific research in pursuit of answers to four basic biological questions, The Land Institute is ready to formally state that our country *can* build an agriculture based on a fundamentally different paradigm than the one humans have featured for the last eight to ten thousand years." Never losing his sly farm-boy humor, Jackson waited for the applause to break and then added, "And not only that, but we think it just might solve all manner of marital problems and end sin and death as we know it." Although the room roared, there was no mistaking the seriousness of what Jackson and his friends had accomplished.

If the eroding Breadbasket is to be transformed by the work at The Land Institute, it will have sweeping repercussions. But our Breadbasket is only one small part of the world's agricultural land. What Piper and Jackson and the rest would never dream of doing is importing prairie agriculture everywhere. The natural systems farm, designed in nature's image, would not look the same in all corners of the world, because ecosystems differ so drastically across the globe. "Take the difference between tropical rain forests and prairies," says Jackson. "In the moist jungle, where water can be too abundant, you want water purgers—plants that can give off water vapor quickly. In the droughty plains, you want water hoarders."

In short, the *genius loci*—"genius of the place"—should dictate the best agricultural system, given the local plant community, climate, soil type, and culture.

What *can* be imported from The Land Institute, Jackson says, is its methodology—its approach to learning a native system, intuiting its "rules," and then slowly trying to raise a stable community of crops that mimics the structure and performs the functions of the wild one. As the following stories will show, the investigation is already under way.

RIPENING PROOF AROUND THE WORLD

"Do Nothing" Farming in Japan

Fifty years ago, when Wes Jackson was a boy weeding his family's farm, a young man in Japan named Masanobu Fukuoka took a walk that would change his life. As he strolled along a rural road, he spotted a rice plant in a ditch, a volunteer growing not from a clean slate of soil but from a tangle of fallen rice stalks. Fukuoka was impressed by the plant's vigor and by the fact that it was up earlier than those in all the surrounding cultivated fields. He took it to be the whisper of a secret revealed to him.

Over the years, Fukuoka would turn this secret into a system he calls "do nothing" farming because it requires almost no labor on his part, and yet his yields are among the highest in Japan. His recipe, fine-tuned through trial and error, mimics nature's trick of succession and soil covering. In early October, Fukuoka hand-sows clover seeds into his standing rice crop. Shortly after that, he sows seeds of rye and barley into the rice. (He coats the seeds with clay so they won't be eaten by birds.) When the rice is ready for harvest, he cuts it, threshes it, and then throws the straw back over the field. By this time, clover is already well established, helping to smother weeds and fix nitrogen in the soil. Through the tangle of clover and straw, rye and barley burst up and begin their climb toward the sun. Just before he harvests the rye and barley, he starts the cycle again, tossing in rice seeds to start their protected ascent. On and on the cycle goes, self-fertilizing and self-cultivating. In this way rice and winter grains can be grown in the same field for many years without diminishing soil fertility.

The neighboring farmers are curious. Whereas they spend their days cultivating, weeding, and fertilizing, Fukuoka lets the straw and clover do the work. Instead of flooding his fields throughout the season, Fukuoka uses only a brief dousing of water to head off weed germination. After that he drains the fields and then worries about nothing, except an occasional mowing of the paths between fields. On a quarter acre, he will reap twenty-two bushels of rice and twenty-two bushels of winter grains. That's enough to feed five to ten people, yet it takes only one or two people a few days of work to hand-sow and harvest the crop.

Natural farming has spread throughout Japan and is being used on about 1 million acres in China. People from around the world now visit Fukuoka's farm to learn both farming techniques and philosophies. The allure of this system is that the same piece of ground can be used without being used up, and yields can be consistently good. Instead of pouring money and energy into the farm in the form of petroleum-based inputs, most of the investment is made up front—in the farm's design.

"It took me thirty years to develop such simplicity," says Fukuoka. Instead of working harder, he whittled away unnecessary agricultural practices one by one, asking what he could *stop* doing rather than what he could do. Forsaking reliance on human cleverness, he joined in alliance with nature's wisdom. As he says in his book, *One Straw Revolution*, "This method completely contradicts modern agricultural techniques. It throws scientific and traditional farming know-how right out the window. With this kind of farming, which uses no machines, no prepared fertilizer, and no chemicals, it is possible to attain a harvest equal to or greater than that of the average Japanese farm. The proof is ripening right before your eyes."

Permaculture Down Under

When ecosystems are efficient and stable, they don't require as much work as those kept in the vulnerable first stage of succession. Australian ecologist Bill Mollison, like Wes Jackson, advocates keeping some crops on the land for many years, to bring farming as close as it can come to nature's efficiency.

For years, Mollison has worked on perfecting a system whereby small-scale farmers would set up a low-maintenance garden, a woodland, and an animal and fish farm and then become self-sufficient—fed, clothed, and powered by local resources that are literally right at hand. Designing *with* nature's wisdom is at the core of this farming philosophy, which is called permaculture, for permanent agriculture. In permaculture, you ask not what you can wring from the land, but what the land has to offer. You roll with the weaknesses and the strengths of your acreage, and in this spirit of cooperation, says Mollison, the land yields generously without depletion and without inordinate amounts of body work from you. The most laborious part of permaculture is designing the system to be self-supporting.

The idea is to lay out crops so that those you visit most frequently are close by your dwelling (Mollison calls it edible landscaping) and those that require less vigilance are set out in concentric circles farther from the house. Everywhere, there are plants in two- or three-canopy schemes, that is, shrubs shaded by small trees, which are shaded by larger trees. Animals graze beneath all three canopies. Dips and furrows in the land are used to cache rainwater and to irrigate automatically. Wherever possible, permaculturists invite external forces such as wind or flooding to actually help do the work. They build windmills, for instance, or plant crops on floodplains, where they can enjoy a yearly pulse of alluvial sediment.

Choosing synergistic planting arrangements—using "companion plants" to complement and bring out the best in one another—is key to a successful agriscape. To maximize these beneficial unions, the permaculturist creates a lot of edge—transition zones between two habitats that are notoriously full of life and interaction. Mollison is also fond of using interactions between animals in place of high-energy inputs or machinery. One example is a greenhouse/chicken coop where plants are stacked on stair-stepped benches. The chickens roost on the benches at night, enjoying the warmth left over from the day's solar radiation. They add to the heat with their own bodies, helping the plants survive the frosty dawns. In the morning, when the greenhouse becomes too hot, the chickens move into the forest for grazing. As they search for nuts and acorns shed by the planted trees, they comb the ground like rakes, aerating and manuring the soil while snatching up tree pests. Humans eat the eggs and eventually the flesh of these chickens, but in the meantime, they enjoy their services as cultivators, pest controllers, greenhouse heaters, and self-fed fertilizers.

Mollison learned this ballet of efficiency firsthand when he worked in the forests of Australia in the late sixties. As a researcher, he was trained to describe the biological world and leave it at that. But Mollison took the next step that is crucial in bio-mimicry: He saw lessons for streamlined living emerging from the forest and vowed to apply them to a new kind of agriculture. Today in Australia many farms are now working according to the permaculture principles he has popularized, and an international permaculture institute, with branches throughout the world, is training people to disseminate the technique. By mirroring nature's most stable and productive communities, and then living right in the middle of them, Mollison believes, human communities can begin to participate in their beauty, harmony, and Earth-sheltering productivity.

New Alchemy Farm on Cape Cod

Another example of ecoculture sprouting in place of agriculture can be found on Cape Cod, at the offices of two of the country's most innovative bioneers, John and Nancy Todd. They formed the New Alchemy Institute in 1969 to design living spaces and food producing systems that would use nature as a model. The forest-in-succession was the conceptual guide for their totally self-sustained farm.

"Conceptually our farm begins at the bottom of the numerous fish ponds, and extends upward through the water to the ground cover formed by the vegetable and forage crop zone where livestock graze. It then rises through the shrub layer to the canopy formed by the trees that produce fruit, nuts, timber, and fodder crops. Following this plan we are hoping to maintain the farm in a dynamic state of ongoing productivity while it continues to evolve ecologically in the direction of a forest," Todd writes in his 1994 book, *From Ecocities to Living Machines*. Like Mollison's permaculture, New Alchemy's farm is designed so that every living component has a multiple function—shading and fertilizing, for instance, as well as yielding an edible harvest. Wherever possible, the work of machines (and, by extension, humans) is replaced by the work of biological organisms or systems.

One of the Todds' inspirations was Javan farms in Indonesia, where unconventional (to us, anyway) agriculture has thrived for centuries. The Javanese farm is nature in miniature, and it shows the restorative processes of planned succession. "Successional or ecological agriculture differs from ordinary farming in that it adapts to changes over time. In early phases, annual crops and fish ponds might dominate the landscape, but as the landscape grows and matures, a third dimension develops as tree crops and livestock come into their own. The key is to mirror the natural tendency of succession which, over time, creates ecosystems that are effective and stable utilizers of space, energy, and biotic elements."

Three-Story Farming in Costa Rica

Succession is also at the heart of a Costa Rican version of Natural Systems Agriculture. The tropical forests here are paradises—cornucopias of irrepressible vegetation and edible foods ripening under a natural heat lamp and mister. It's therefore all the more ironic, and perhaps telling, that jungles like this have made such poor sites for growing conventional crops. The first few years after clear-cutting and/or burning a primary forest, crop yields are good, but then they drop precipitously. It makes sense if you realize that the same force that creates the jungle—deluges of rain—can also leach nutrients from unprotected jungle soil after clearing, when there are no plants around to soak up water. Crop harvests also remove even more nutrients from the site. After a few years of this nutrient extortion, the soil quickly tires.

Natural clearings in the jungle meet an entirely different fate. They are quickly revegetated by a parade of species that take over one after another, sinking roots, spreading canopies, shedding leaves, and restoring fertility to the site. Nutrients in the system are kept in play in the green growing biomass—nutrients "on the stump."

John J. Ewel, a botany professor at the University of Florida, Gainesville, hypothesized that if you could simulate a natural regrowth of jungle using domestic crops as stand-ins for the wild species, you could achieve the same fertility-building phenomenon and actually improve the system rather than deplete it. The trick is to start with crops that mimic the first successional stage (grasses and legumes), and then add crops that mimic the next stage (perennial shrubs), all the way up to the larger trees—nut crops, for instance.

To test their hypothesis, Jack Ewel and colleague Corey Berish cleared two plots in Costa Rica, letting them naturally reseed to jungle. In one of the plots, every time a jungle plant sprouted, they would dig it up and replace it with a human food crop that had the same physical form. Annual for annual, herbaceous perennial for herbaceous perennial, tree for tree, vine for vine—it was as if nature were guiding the hands of the agronomists. The parade of volunteers to the natural system (*Heliconia* species, cucurbitaceous vines, *Ipomoea* species, legume vines, shrubs, grasses, and small trees) were replaced by plantain, squash varieties, yam, and (by the second or third year) fast-growing nut, fruit, and timber trees such as Brazil nuts, peach, palm, and rosewood.

This domestic jungle of crops looked and behaved like the real jungle in the plot next door. Both plots had similar fine root surface area and identical soil fertility. The researchers also put in two control plots: a bare soil plot and a plot planted in a rotating monoculture—maize and beans followed by cassava, followed by a timber crop. While the bare soil and the rotating monoculture lost their nutrients very rapidly, the "domestic jungle" remained fertile.

Several years before Ewel's paper came out, British permaculturalist Robert Hart also published some concrete recommendations for cropping systems that would mimic the jungle ecosystem. They included cassava, banana, coconut, cacao, rubber, and lumber crops such as *Cordia* species and *Swietenia* species. At the end of its succession, Hart's cropping system would be a three-tiered canopy, mimicking the structure of the jungle as well as its nutrient cycling, natural pest control, and water-purging function. The trick to keeping the soil fertile, says Hart, is to choose perennial crops with lots of leaves and roots, so they can protect the soil from hard rains, store nutrients in biomass, and put organic matter back into the soil when they shed. Hart also found it important to use plants that form symbiotic associations, as well as deep-rooted plants that pumped nutrients from different depths of the soil. In this way, the ground was kept continually covered, yields were provided throughout the year, and each set of new crops prepared the soil physically and even chemically for the next stage. Once the succession progressed to tree crops, farmers could selectively harvest timber and burn the perennials every few years to start the cycle again. Besides supporting local farmers, this sustained usefulness may also help to slow the relentless clearing of primary jungle.

The New England Hardwood Forest

Radical as it seems now, mirroring ecosystems is not a new concept. Sir Alfred Howard, whom many credit with the invention of organic agriculture, talked about farming to fit the land in his 1943 book, *An Agricultural Testament*, as did J. Russell Smith, in his 1953 book, *Tree Crops: A Permanent Agriculture*. Smith wanted to see eastern hillsides replanted with tree crops, which seemed to suit the hills better than the erosion-causing row crops planted after the great green wall of New World forest was torn down.

Smith looked to the eastern deciduous forest as a model of diversity and stability. He described the great number of niches provided by the various tree-canopy levels as well as shrubby and herbaceous understories. Thanks to the diversity, he wrote, pests are kept under control and birds and browsing animals are given many places to make a living. Fine fibrous roots of woody understory plants act like a prairie's sod to hold soil and retain nutrients. Fallen leaves and debris are slowly and steadily recycled into new plant life, preventing leaching and downslope loss of critical nutrients. The organic litter also encourages the growth of mycorrhiza—fungi that form associations with roots and further extend their water-searching power. Every now and then, wind or disease or lightning takes out a tree, creating a gap where succession and renewal can begin again.

Early agriculture on these soils, practiced by Native Americans, was also successional in nature. The tribes practiced small-patch agriculture, raising beans, squash, corn, and tobacco on twenty- to two-hundred-acre plots. After eight to ten years, the native farmers would move on and allow the land to lie fallow. In the twenty-year hiatus before the farmers returned, succession would resume and fertility would be restored. This shifting method required tribes to be nomadic, but it mimicked the natural forest dynamism by creating small patches that were allowed to revert to forest.

In his book, Smith bemoaned the loss of soils and productivity that occurred when white settlers began to farm more permanently on these sites, deforesting hillsides and planting row crops. The farming didn't fit the land, he claimed. Instead, he proposed planting structural analogues—nut- and fruit-bearing trees as the only fitting crops for forest-growing land. One scheme that bore out his dream was a farm of honey locust trees (which bore seed crops) with an understory of Chinese bush clover (a perennial legume suitable for grazing and haying). This system yielded crops and supported animals, all with minimal labor, low management costs, and good weed control. He reported returns of 4,500 pounds of hay per acre per year, 2,920 pounds of honey locust nuts per acre per year on average, with a peak of 8,750 pounds of nuts per acre in eight-year-old trees.

The features that made the hardwood forest sustainable in the wild were repeated here: a tree crop in the overstory, a stable understory to protect the soil and retain nutrients, a biological nitrogen source, and a grazing or browsing animal component. Unfortunately, Smith's recommendations fell largely on deaf ears when his report was first issued. The fact that his work has been republished by Island Press recently, with a foreword by Wendell Berry, is a hopeful sign that the idea of nature-based farming is sprouting once again.

The Desert Southwest

Where prairies and forests fear to tread, the model for farming is an unlikely one—the scrubby, spiny desert of the American Southwest. Across the Sonoran, the Chihuahua, and the Mojave, rainfall is erratic and strongly seasonal, and soils may vary every few feet. These uneven conditions lead to a patchiness of vegetation—plants cluster in fertile alluvial fans, while on more barren stretches, they space themselves out to get all the water they can. Besides dividing up the space, they also divide up the season. Many species bloom and set seed only when water is available, becoming dormant as the summer blisters on.

These strategies, which allow plants to take advantage of ephemeral resources and to endure long dry spells, were mirrored in the farming methods of original peoples who flourished here for thousands of years. The Papago and Cocopa peoples continue to live here, gathering their foods from both wild plants and cultivated desert plants and legumes, all of which are native to the place, thus adapted to making the most of limited resources. Ethnobotanist Gary Paul Nabhan made readers aware of their agricultural practices in his book *Gathering the Desert*.

To the extent possible, writes Nabhan, the Papago synchronize their agriculture with the local seasonal clock. Planting, for instance, is timed to the emergence of desert annuals—right before or after nourishing rains. By planting only on flood-watered alluvial fans, they avoid having to intensively irrigate, which in that climate of excessive evaporation would leave poisoning salt in the upper registers of the soils. Besides annuals, the Papago also sow succulents, grasses, and woody plants for food and fiber. Interspersed with the crops are wild mesquite trees, left in the fields because they can fix nitrogen and gather deeply stored soil nutrients. Long before agronomists knew why this companion planting worked, the Papago were practicing it, having taken their cue from the "genius of the place."

Rodale's Regenerative Agriculture

No talk of organic agriculture would be complete without a mention of the Rodale family, whose legacy includes the Rodale Press as well as such publications as *Organic Gardening Magazine*, *New Farm*, and *Prevention*, a magazine devoted to health issues. Like Mollison's permaculture, Rodale's "regenerative agriculture" uses biological structuring to increase the efficiency of nutrient and energy flows so that low-energy inputs are leveraged into high productivity. Succession is also used strategically. Crops are carefully chosen to change the soil flora and fauna in a way that anticipates the needs of the next crop. For instance, practitioners may plant a crop that causes the weed community to shift toward species that are not a problem for the next crop. Or they might emphasize nitrogen and soil-carbon buildup in one part of the rotation cycle to increase the productivity of subsequent crops. Finally, researchers at Rodale have spent some time, as Jackson has, looking for perennial replacements for annuals such as wheat, rice, oats, barley.

Letting the Cows Out in the Midwest

Crop growers are not the only ones caught in the box canyon of industrial farming. For years now, dairy farmers in the upper Midwest have been cutting hay with machines instead of letting the cows graze it. They've been tractoring the fifty-pound bales into their artificially lighted and heated suction-milking barns.

Now all that is changing. Dairy farmers are opening the doors to both their minds and their barns in a nature-based movement called "grass farming." Dairy farmers who have switched to grass farming are now letting their cows munch at least three of the five hayings in the field. They report that they enjoy the work of bringing the cows to their food rather than the other way around. Grass farmers also find that their cows are healthier and their bills are slimmer. Manure in the fields means they can pare back their fertilizer bills, and because they hay with machinery only twice, they also save money on fuel and machine wear.

After a few years, many of the farmers are shifting to an even more natural cycle. Instead of milking their cattle year-round, they "dry them off" during the winter, so they can calve all at once in April and be ready to go back to the grass in the spring. This dry-off allows the grass farmers to do what had been unthinkable in the old system—take a vacation.

The term *grass farming* signals a change in how the farmers see themselves. "They consider themselves solar harvesters now—turning sunlight into grass and then into meat and milk," says Stephanie Rittmann, who wrote her 1994 master's thesis (University of Wisconsin-Madison) on the swelling movement and how and why it is spreading. "What's interesting to me is what grass farming has done to community life in the rural Midwest," says Rittmann. "Because these farmers are trying something completely new, they are all at the entry level in terms of know-how. No one is a complete expert on managing grass pasture for their herds. In fact, one of their only guides is a book called *Grass Productivity*, written by French agronomist André Voisin in 1959. Beyond this, they turn to one another for advice, and have formed a long-distance support community." They visit one another's farms periodically to share what they're learning and they produce a monthly newspaper called *The Stockman Grass Grower* that is filled with candid dialogue between producers.

To grow good pasture, grass farmers face many of the same challenges that prairie restorers face. They begin with an alfalfa field, then sow in about four species of grass. As the years wind on, wild plants infiltrate, some that the farmers have never seen before. As Rittmann says, they are watching succession on their lands and comparing notes, learning what the land might have looked like before the plow.

They are also using new ways to assess the health of their pasture, and here's where the farmer becomes a naturalist. One man was at first puzzled and then absolutely thrilled to hear a strange crackling noise in his fields—the sound of hundreds of thousands of earthworm holes opening back up after a rain. "Finally I realized: That's what a healthy pasture is supposed to sound like," he told Rittmann. Another farmer said it took three years of grass farming before he finally heard birdsong returning to his pastures. Now he counts and catalogs the bird diversity around his pastures as a way of assessing their health. Other grass farmers look to cowpies—a cowpie in a healthy microfauna and microflora should break down in three weeks' time in midsummer. If it's around any longer, the farmers tell Rittmann, they start to worry.

"What they are doing is learning how to read nature instead of simply relying on the word of a pesticide salesman," she says. "I tell them they are starting to act like ecologists, and they just shake their heads and smile. 'Nah. It's just farming,' they tell me." Smart farming.

RADICAL DEPARTURES: HOW DO WE GET OFF THE TREADMILL?

The spread of the grass-farming idea should be studied carefully for clues. Just how does an idea "take" in the imagination of a group that is culturally and economically entrenched in a certain way of doing things? How will The Land Institute sell its idea to farmers who are already treading water as fast as they can just to keep up? How do you spring the mind free from its fears?

Wes Jackson is well aware of all the things our minds have to overcome. For starters, he describes the mind shaped by reductionist science, the American experience, evolution, and affluence. "We have convinced ourselves that the universe is comprehensible in small separate pieces, that there is always more frontier, that any new technology is adaptive, and that there are, as author Wallace Stegner says, 'things once possessed that cannot be done without.' " This mind conditioning makes it tough for us to think of the whole, respect nature's limits, or pass up what technology promises, be it convenience, wealth, power, predictability, or cheap food. How, then, will the Breadbasket become a domestic prairie?

"Not all at once," says Piper. "We'll begin by offering Natural Systems Agriculture as an alternative on Conservation Reserve Program lands." The Conservation Reserve Program (CRP) was begun in 1985 to heal the hemorrhaging scars from the fencerow-to-fencerow era. Farmers are paid an average of $48 an acre to retire their erodible lands and plant them to perennial grass. So far, 36.5 million acres have been planted through CRP (if you add the land set aside in previous programs, it comes to over 100 million acres of grassy slopes). Unfortunately, many of those acres were planted in exotic grasses that are of limited use to wildlife and offer "focused" farmers (who abandoned their livestock) no way to make an income.

Perennial polycultures on those same lands would offer farmers an income in addition to holding down their soils. They could collect their income in one of three ways. They could hay the domestic prairies, harvest the seed for human consumption, or, if they have livestock, simply graze them. This way, the income would come back to the farmer, instead of being shipped off to the manufacturers of pesticides and fertilizers. The time is right for this sort of transition, Piper feels, because the CRP is due to sunset soon, and it may not be renewed. In a survey conducted by the Ohio Soil and Water Conservation Association, 63 percent of farmers said they were planning, for economic reasons, to plow up their CRP lands if subsidies dry up. Perhaps, if they hear about The Land Institute's work, they can hold out for a whole new idea—that of *healing the soil while growing food.* To a culture accustomed to causing damage, that sounds sweet to the ear.

But perennial polycultures won't take over the whole farm landscape, predicts Piper. There are some noneroding bottomlands that are perfectly suitable for planting in row crops—under an organic regime, of course. "But that's only one eighth of our cropland," he says. "The other seven eighths consists of erodible soils and sloping ground, and it suffers when row-cropped. On these lands, Natural Systems Agriculture makes more ecological sense." But will it make sense to farmers?

Ultimately, the strongest persuader is likely to be changing economic conditions. When the way farmers (or anyone else, for that matter) have been doing things becomes economically uncomfortable, they will be eager to try something new. This may happen when fossil fuels begin to run out, making farm inputs such as gasoline, fertilizer, and pesticide prohibitively expensive. When that time comes, we'll do what any species does under the pressure of change. We'll start shopping around for alternatives and adopt the most creative one, jumping to the next evolutionary level.

At The Land, they call this next level "the sunshine future." If you ask, staff members will indulge in a dream of what a farm in the sunshine future would look like: The new Breadbasket farmers would tend domestic prairies—seed-producing perennial mixtures—which would build soil instead of squandering it. Because of its chemical diversity, the farm would naturally protect itself from most pests, tamping down populations before they reach epidemic levels. Weeds would be managed by the chemical interaction of plants and by shading. Nutrients would be held in the soil instead of leaching out. Pesticide and fertilizer use would be minimal, maintenance light, and plantings infrequent. A farmer could start over with a new crop of perennials every three to five years, but would do so by choice, not by necessity.

Livestock would also require less babying. Beef cattle are now being bred with buffalo, for instance, to produce animals with tougher hides, like barns on their backs. These beefalo could be left outside in winter, obviating the need for lumber to build protective structures. Throughout the year, they could be moved from one polyculture to another in a rhythm that does not jeopardize flowering and seed set. Their wastes would contribute to the crumb structure of the soil, which, along with root action, allows the sod to wick moisture in and allocate it slowly. More water-holding capacity would mean less call for irrigation. It might even encourage springs to reopen as underground reserves are recharged.

Until we are farming in the sunshine future, Jackson has written, groups like The Land Institute are, in the Buddhist sense, "making a path and walking on it." Research, economics, and community will all play a role in how successful their journey is. The following is an attempt at an itinerary.

Consulting the Genius of the Place: Research

Wes Jackson compares the typical agricultural researcher to the proverbial drunkard who is looking for his lost keys under the streetlight. When asked why he is looking here when the keys were lost up the street, he replies that the light is better here. In like fashion, our research institutions have searched for agricultural advances where the money is—in the glare of industrial farming. Taxpayers foot the bill in the form of appropriations to USDA research and in the form of 20 percent investment credits to new private research facilities.

What are we paying for? Right now, the bulk of research helps to shore up the system of farming that is already in place. Most disease dollars, for instance, are spent on diseases that afflict only crops grown in continuous culture, a system we know is anathema to soil fertility. Instead of investigating markets for alternative crops (those that can be grown in rotation), our economists continue to invent new markets for the big input-hungry four: wheat, corn, rye, and soybeans. And, of course, a lot of money goes toward breeding crops that will withstand chemicals.

"Where are our values?" asks Gary Comstock, a philosopher at Iowa State University. "Now that atrazine has turned up in the wells of some farm families, 2,4-D has been linked with non-Hodgkin's lymphoma in farmers, and Alachlor, the most heavily used herbicide on corn, is suspected to be a carcinogen, why are land-grant universities doing research to find crops that can be grown in the presence of stronger doses of it?"

If research is a form of social planning, as Chuck Hassebrook of the Center for Rural Affairs says, what does this say about where we want to go as a society? Instead of pledging allegiance to a method of farming that we know destroys land and people, shouldn't we be tackling the problems of getting crops to grow the way we *want* them to grow—in polyculture and rotations, for instance? Shouldn't we be taking nature's advice and giving farmers the tools they need to farm sustainably, rather than giving chemical companies bigger needles to poison us with? The Land Institute had been striving to *keep* arable land arable for twenty years now, with negligible federal assistance. It was time, they decided, to knock on government's door and bring research spending in line with society's hopes for the future.

Wes Jackson had been waiting for just the right moment. When The Land staff members had scored five articles in prestigious scientific journals, he put on his meeting clothes and went to Kansas congressman Pat Roberts, who was the Agriculture Committee chair at the time. Jackson laid out a plan for several sites around the country that would be centers for Natural Systems Agriculture. This network would take this agricultural Kitty Hawk and put it through fifteen to twenty-five years of wind-tunnel tests in different climatic regimes. Look here, said Jackson to Roberts, isn't the marriage of ecology and agriculture the sort of research that government should support? The congressman answered with a question. "What does the university think about all this?"

So Jackson trudged back to Kansas and received a glowing endorsement from Kansas State University. After many more visits and yeoman phone work from a man who would rather be threshing gamagrass, the committee told Jackson: "We'll look into it." Those four simple words had never before crossed the lips of conventional agricultural researchers. Nor had statements like the KSU mission statement, which admits, "A new agricultural research paradigm is needed." The people sitting near me at the Politics of Sustainable Agriculture conference in Eugene, Oregon, were understandably shocked, then, when Jackson announced, "On September twenty-eighth, 1995, the conference committee of both houses agreed to include language in the 1995 Farm Bill that essentially instructs the Secretary of Agriculture to investigate and support Natural Systems Agriculture."

A spontaneous cheer combusted in that room, and we gave Wes Jackson a standing ovation.

Setting Up the Books: Energetics

After we all sat down, Jackson started rhapsodizing about his latest passion. He's been telling everyone who will listen that accounting is going to be the most exciting profession of the new century. Accounting. We laugh, and then he explains that ecologists are a breed of accountant. One of the ecologist's primary tools for measuring and describing the sustainability of ecosystems is to draw a circle around the system, tote up all the inputs and outputs, and then analyze the energy cycles inside the circle. Again and again, in terms of energetics, natural systems miraculously "pencil out"—they remain viable without drawing down their resources. If we are to switch to a more natural agriculture, says Jackson, our systems must also pencil out, in at least two ways: 1) Economically, they must sustain farmers and their communities, and 2) ecologically, they must pay their own energy bills and not draw down the resources of the local landscape or the planet.

The surest path to sustainable farming, says Jackson, is to make sure the lion's share of rewards runs to the farmer and the landscape. Marty Strange, codirector of the Center for Rural Affairs, puts it this way: "To be sustainable, agriculture must be organized economically and financially so that those who use the land will benefit from using it well and so that society will hold them accountable for their failure to do so." For society, it may mean changing economic policies so that our well-being, including our environmental well-being, is reflected in the gross national product. It may mean pricing food commodities to reflect their true costs. It may mean eliminating some of the tax breaks that encourage the substitution of capital for labor and essentially subsidize irrational farm expansion and over-production. In their place, says Strange, we should design policies that give a hand to farmers who are more likely to treat the land well—those on owner-operated, family-held, and internally financed farms. To stay viable, these farms must ultimately break the unhealthy coupling they now have with the petroleum and chemical industries.

Whenever you break the cycle of dependency, you inevitably hear the anguished moans of the addict in withdrawal. Without large farms and fossil-fuel amendments, will we still be able to feed our-selves? Will we be able to feed the world? Piper's answer to the first question is yes. "Although yields may not be as high, we ought to be able to feed ourselves and then some. Consider that we have had a grain surplus every year since the thirties in this country, and that *eighty percent of our grain is not fed to people but to livestock*." (We feed cows grain to "finish them," that is, to marble their meat with the fat that clogs American arteries.) Piper feels there's obviously some slack to be taken up here. As for feeding the world, he says, "Maybe the better goal would be to enable the world to feed itself." But that's another subject.

The point is that the sanctity of seeking higher yields—the ag-ronomic equivalent of the search for gold—makes it virtual heresy to drop down to more realistic yields, to what the land will support over time. The Land realized that in order to defend the yields of perennial polyculture against those of conventional monocultures, it would have to somehow level the playing field. Piper puts it this way: "If we said to a wheat field, 'Sponsor your own fertility, grow without pesticides or diesel fuel for traction,' then what would the yields be? Once you take away the crutches of industrial farming, would it be more economical to grow perennial polycultures or con-ventional crops?"

Piper answers his own question cautiously: "The perennial po-lyculture scheme—planting a prairie that stays put—is designed for low inputs. Cutting down on maintenance, fertilizer, and pesticides is bound to save money, perhaps enough to make this form of farm-ing as competitive as its fuel-dependent cousin." Jackson is less cir-cumspect: "Perennial polycultures would beat the pants off conventional crops grown in a sustainable way. Period. But now we need the data to prove it."

Once again, The Land staff went to the literature, and once again, they were disappointed. There were studies on organic (pesticide-free) farms, but none on organic farms that also grew their crops with-out fertilizer and without diesel fuel. After twenty years, a lack of published data had come to look more like a red cape than a stop sign to this group. So in 1991, they pawed at the ground a few times and began the Sunshine Farm project: one hundred and fifty acres, con-ventional crops, tractors that use vegetable oil for fuel, photovoltaic panels for electricity, draft horses for some field operations, longhorn cattle for manure and meat, hens that turn compost (then turn a profit with eggs), and broilers that forage in alfalfa. In all, a demonstration farm where biological and solar energy are expected to pay the bills.

"Sunshine Farm is really one big accounting project," says Marty Bender, the towheaded energy accountant of the farm. Over coffee, he stokes up his computer and shows me a giant database. "We draw a big circle around the farm in our minds, and then we count up everything that comes into the farm and everything that goes out, using techniques that are very similar to what ecologists use to describe the energetics of an ecosystem. We literally measure the size, weight, and amount of everything—every fencepost, every galvanized gate, every foot of chicken wire, every plastic pail. We figure out how much energy it takes society to make that product, and then we record it in kilocalories."

To track labor, Bender has devised a taxonomy of tasks done on the farm—weeding, fence-mending, broiler feeding, and so on, so that every finger that is lifted can be accounted for in kilocalories. A trip to the store for ten-penny nails takes fuel, labor, and the energy society expended to manufacture the nails—all debits against the farm. In turn, everything the farm produces—all crops, livestock, biofuels, and so on—is recorded as an asset. The trick is to balance the budget so the farm is not a drain on the planet.

Bender's energy estimations come from an enormous literature search. When you are with him, he frequently dashes to his wall of filing cabinets to grab one of the hundreds of articles he has gathered, with titles like "The Embodied Energy Content of Polyethylene Pipe." Each article is covered with the furiously scribbled notes (sometimes corrections) that are his trademark, an artifact of his brilliance.

"There's nothing else even remotely as thorough as the Sunshine Farm database," Bender tells me. "So far, we've logged over twenty-seven hundred transactions and we're not even half through. Keeping the ecological books this way will tell us whether a farm can run on sunlight and keep its books balanced—that is, pay all its own bills without going into debt to the larger environment." In other words, can the farm itself produce enough food over time to support the human and animal labor, provide fuel for its machines, and manure for its fields? Can it do all this and grow crops that will reimburse society for the energy embedded in material off-farm purchases? Answers like these will tell us what agriculture really costs, and perhaps, says Bender, suggest a more accurate, long-term cost for what we eat. "That's real important."

As we talk, Jack Worman, the farm manager, comes in, wearing a ten-gallon hat that makes me remember how far west we are in Kansas. The creases in his face, if you counted them, might tell you something about the drought cycles in this part of the world. With impeccable cowboy manners, he touches his hat, apologizes for interrupting, then consults with Bender, not about the chickens or the crops but about the kilowatt meter that monitors the solar-panel array. This is not your ordinary farm operation, I conclude, at least not yet.

Right livelihood might be voluntary today, but The Land Institute predicts that someday it will be mandatory. When fossil fuel runs out or becomes too expensive, people will have to do sunshine farming. In the meantime, Jackson hopes the Sunshine Farm will not be an isolated experiment. He writes, "Until we have the physical manifestation of sustainable livelihoods demonstrated in enough places, we are going to continue the folly. So the good examples, whether they are the good examples among organic farmers, or the good examples among research efforts, or just the good examples of ordinary right livelihood, give us a standard." Nature as measure.

Becoming Native to This Place: Community

None of this is going to happen in isolation. If we want to weave the ecological paradigm into our research and our economy, we need to bring people back to farm country. Nature teaches us that ecosystems are made up of habitat specialists—local experts who know how to work the system. One hundred and fifty years of farming the American plains has also resulted in an accumulation of local knowledge. People have learned how to time plantings, how to read the weather, and what to expect from soils, insects, diseases, and each other.

The problem is that with the rapid depopulation of the countryside, this knowledge has been disappearing. At this point, only 1 percent of the U.S. population is growing our food, and that figure is falling. Half of all farmland is owned by nonfarmers; only seven companies run 50 percent of the farms. As Wendell Berry observes, no one bemoans the fact that a farm Grange is closing for lack of members; in fact, we are more scandalized by the loss of indigenous rain forest cultures than we are by the loss of American rural cultures.

Jackson notes that this loss of farmers is not the first but the second wave of loss. Native Americans were the repository of a much longer cultural history, but we've already moved them off the land. Now we're on to our second wave of "surplus" people. If Natural Systems Agriculture is to be successful, insists Jackson, we need a homecoming of people willing to "become native to their place," tuning their senses to local conditions, and farming the land in a way that will last. You can't expect people to buy small farms and re-populate the countryside, however, unless they are able to make a living and a fulfilling life for themselves far from town. That will require a restoration of community, says Jackson, not because it's nostalgic but because "more eyes per acre" is a practical necessity.

Moved by this belief, Jackson decided to learn what he could about human communities in rural areas. "We asked the question, why shouldn't human communities run on sunlight and recycle materials the way natural communities do? Why can't our home places be sustainable instead of simply being quarries to be mined by the extractive economy and then abandoned? After all, native peoples lived here for hundreds of years, in far greater concentrations than we have today in some rural counties. How was it that the land could support them in a sustainable way?"

To answer that, Jackson decided to spend some time with the remaining inhabitants of one of the quarries—the fifty-some-odd townspeople of Matfield Green in Chase County, Kansas (the site of William Least Heat-Moon's *PrairyErth*). During the late 1980s and early 1990s, he bought the abandoned elementary school (a beautiful, ten-thousand-square-foot brick structure built in 1938) for $5,000, the hardware store for $1,000, and with some friends, seven abandoned houses (including one he plans to retire in) for less than $4,000. His nephew bought the bank for $500, and The Land Institute bought the high school gym for $4,000. Friends and employees of The Land have since begun to move into town, restoring their homes with used lumber and other renewable technologies and transforming the school into an education center and conference space for artists, scholars, and teachers interested in becoming native to their places.

Emily Hunter is the smart and passionate coordinator of the Matfield Green Project. "Forget Paris," says Hunter. "The cultural capacity to live sustainably resides right here, in the residents of Matfield Green, those people who decided to stay after the boom-and-bust and figured out how. We realize that if we want to join them in this beautiful tallgrass prairie, we can't repeat the mistakes of the extractors. We have to live in a way that doesn't spend the ecological capital of the Flint Hills region. Instead we're asking, what is the wisdom being expressed today by this tough, rooted town? It's been pruned and burned back by the fossil fuel economy, and maybe it's back to rootstock. What can we safely graft onto that? How can we create patterns of sustainability together? The people of Matfield—like Evie Mae Reidel who knows what phase of the moon is best for planting potatoes—can help us discover those patterns. With their help, we can teach other homecomers."

For now, the learning takes place over coffee at the restored lumberyard and in meetings at the renovated school. Each month, the Tallgrass Prairie Producers, a cooperative devoted to raising prairie-fed cattle, gathers to strategize in one of the old, high-ceilinged classrooms. During the summer, workshops will be held here for teachers who are designing a place-based curriculum for rural schoolkids.

In the meantime, staff from The Land are conducting an environmental history of the area to see decade by decade how land use has changed. This is the first phase of an ecological community accounting project designed to determine the human carrying capacity of a place. "We know we are in deficit," says Hunter. "Our job is to find out how to be sustained by a place without bankrupting it. Our teachers are the prairie and the people who have been shaped by the prairie for generations."

Jackson says residents here and in similar communities are "the new pioneers, homecomers bent on the most important work for the next century—a massive salvage operation to save the vulnerable but necessary pieces of nature and culture and to keep the good and artful examples before us."

CROSSING INTO THE EDDY

Matfield Green, Sunshine Farm, and other right-living projects around the world are attempts to create counterpoints to the extractive economy, to "keep the good and artful examples before us." I think of them as eddies in a turbulent whitewater river.

An eddy is a pocket of calm water that forms as water passes around a rock, leaves the downstream current, and curls back upstream to form a magic haven in the rock's shadow. It's a place a kayaker can duck into when she needs to rest, take stock, or rescue less maneuverable boats from calamity.

Getting your boat into an eddy is hard work. You must cross the line of tension, the rip between the downstream torrent and the curling upstream flow. It takes some momentum and a vigorous, well-placed paddle brace to pivot across the eddy line and into the sanity of smoother water. In the same way, our transition to sustainability must be a deliberate choice to leave the linear surge of an extractive economy and enter a circulating, renewable one.

Wes Jackson thinks it appropriate that agriculture be the first eddy we enter. He has often called agriculture the Fall, the beginning of our estrangement from nature. "It is fitting then that the healing of culture begin with agriculture," he says. Natural Systems Agriculture is as different from conventional agriculture as the airplane was from the train. It's an evolutionary leap in innovation.

The difference with what we are doing, says Piper of The Land's work, is that no one can immediately cash in on it. After all, when seed companies or chemical companies see a cropping system that needs no seeds or chemicals, they're more likely to fight it than join it. The only logical champions of this revolution are consumers who care about how their food is grown, small independent farmers, and a government that represents them. The transition will start slowly, predicts Jackson—if we're lucky, scattered examples of a circulating renewable economy will appear right alongside the extractive one, and people will suddenly see that they have a choice.

Already people are supporting agriculture that attempts to wean itself from fossil fuels, at least where pesticides and excessive tilling are concerned. The popularity of certified organic foods, food-in-season restaurants, and community supported agriculture (CSA) are a few examples of eddies that are forming in the river. Through CSAs, city dwellers subscribe with a local organic farmer at the beginning of the season, then pick up a bag brimming with fresh produce each week of the summer. The farmer gets the money up front, and the buyer shares in the risk, agreeing to eat whatever crops do well and do without those that fail. In this way, consumers learn to eat with the cycles of the local landscape and have the satisfaction of knowing their food is grown nearby and in conscientious ways.

According to Russell Ubby, director of the Maine Organic Farmers and Gardeners Association, 523 farms in North America are now doing business via this pre-pay share method. Wisconsin has the most, he says, followed by New York and California. The largest of the farms supplies more than two hundred families yearly.

That more people are beginning to care about this aspect of our lives does not surprise me. The idea that food is more than a commodity is deep within us, which makes the thought of a square tomato seem outrageous, or at least distasteful to most of us. We know that the scale of farming should be smaller and more personal—that the land would be better served by stewardship than by massive tractors sporting six TVs. The novelist Joseph Conrad said that there are only a few things that are really important for us to know and that all of us know them. We want our farmers to be breaking off an ear of corn to taste a kernel right before harvest. We instinctively want them to heft the soil, to smell it and know what's wrong or right about it. And I think that instinct comes from our biological urge to survive. It's that visceral common sense that causes a part of you to rejoice when you see crocuses returning and to be revolted when you hear about tons of U.S. topsoil washing into the Gulf of Mexico.

Food is something we have it in our genes to care about, and we have been severed from that caring for too long. If we could once again regard the act of growing food as a sacred, biological act that connects us to all living creatures, perhaps we would clamor for a system of farming that builds communities, maintains balanced pest populations, keeps soil out of rivers, and doesn't traffic in chemicals that are alien to our tissues. Perhaps we'd seek out examples of practical reverence, like those of Wes Jackson and Bill Mollison and Masanobu Fukuoka.

On the surface, these men seem to be tilting at windmills, bucking a strong sea of "how it's always been," and railing at habits acquired ten thousand years ago. In reality, they are the conservatives, secure in the knowledge that their ecomodel is older than agriculture, and that it will be here long after oil-driven agriculture is a memory. This is not really a new fangled thing we are inventing here, insists Jackson. It is just a matter of discovering what is already there and mirroring it.

All in all, I think nature-based agriculture will be nourishing in the best sense of the word—an honest and honorable way to take our place in the food web that connects all life. We have lived too long by hubris, imposing disruptive patterns on the land, squaring the circle. If we as a country, or as a global net of communities, are truly committed to sustainability in all things, agriculture must be first on our agenda, the first meal of the new day. A change this grand will take the cooperative will of all of us, and it will be based on the one characteristic we all share—a primal need to eat. When we begin to insist on nature-based farming (or, as Jackson says, when trendy people in restaurants start whispering, "Do you believe so-and-so is still eating *annuals*?"), then we'll have planted a giant paddle brace against the rapids of environmental disaster. We'll have crossed into the eddy, showing the world, and ourselves, that it can be done.

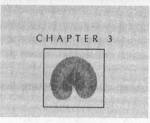

CHAPTER 3

HOW WILL WE HARNESS ENERGY?

LIGHT INTO LIFE: GATHERING ENERGY LIKE A LEAF

"Pond scum" may be a synonym for "primitive," but the tiny organisms that compose it easily beat the human state of the art when it comes to capturing energy from the sun. Some purple bacteria answering to that unflattering description use light energy with almost 95% efficiency—more than four times that of the best man-made solar cells.

—University of Southern California news release,
August 22, 1994

The energy sector in industrialized societies is probably the single largest economic contributor to global environmental degradation.

—EPA's Expert Workshop on Energy and the Environment
July 21, 1992

When I first began dreaming about this book, I would sit at the edge of my pond and watch Montana clouds skate upside down across the water's surface. At night, I'd watch the moon pole-vault up and over. That was before duckweed moved in and stole the big sky show away.

Duckweed is a floating plant with a single round leaf, as thin as paper and no wider than a pencil eraser. It spends its winter *alive* at the bottom of my frozen pond, feeding on its own stored starch. One buzzy May day, it pops up as if arriving for an appointment, and then, to put it mildly, it multiplies. In a matter of weeks, it has stretched a living lid of lime-green leaves across every square inch of water surface. By August, when the leaves of cattails and cotton-woods have grown dark and dusty, duckweed is still exuberantly green, so springtime green that people stop their cars to stare. We thought it was wet paint, they tell me.

En masse, duckweed spreads an impressive solar array—*one* plant, a mere quarter of an inch across, can multiply through the sheer energy of sunlight to cover an area the size of a football field in a couple of months. But there is not just one; there are millions of them. I screen them off; they grow in behind me, like splinters from the Sorcerer's broomstick. This spasm of photosynthesis—sunlight transformed into acres of green tissue before my eyes—is more than just my nemesis. It's a miracle.

That's what most folks thought before the late eighteenth century when scientists began experimenting with leaves to learn "from whence their mysterious nourishment came." This was at a time, remember, when mice were believed to spontaneously arise from piles of rags. Joseph Priestley, an English amateur chemist, mystified the curious when he published the results of his bell jar experiment in 1771. He had sealed a mouse and a candle inside a jar, and the mouse had died, asphyxiated by the "injured air." Miraculously, when Priestley added a mint plant to the mix, he could add a new mouse, and it would live. Vegetation, he told the world, can somehow repair air.

But in the devilish way that photosynthesis research seems to work, Priestley was plagued for years by disappointment when he tried to repeat these experiments. Historians think he must have moved his jar to a darkened corner of his lab, not knowing that light played a role in the release of oxygen from the mint leaves. Mouse after mouse kept passing out. It took eight more years before Dutch physician and chemist Jan Ingenhousz did the same experiment near a sunny window and had a lightbulb of revelation blink on.

The rest is history. We now know that photosynthesis, which means "putting together with light," is the process by which green plants and certain algae and bacteria take carbon dioxide, water, and sunlight and transform them into oxygen and energy-rich sugars. In the meantime, animals like us take that oxygen and those sugars and transform them back into carbon dioxide, water, and energy. Thanks to the sun, mint and mice and men all thrive.

We on this bell jar called Earth are lucky to be so close to such a marvelous explosion happening all day, every day, above our heads. The sun's fusion of hydrogen provides enough light energy to easily supply all our energy needs without burning a drop of oil. If only we had a way to plug in.

So far, we've lived by the grace of green plants, and we owe both our lives and our lifestyles to them. Consider that everything we consume, from a carrot stick to a peppercorn filet, is the product of plants turning sunlight into chemical energy. Our cars, our computers, our Christmas tree lights all feed on photosynthesis as well, because the fossil fuels they use are merely the compressed remains of 600 million years' worth of plants and animals that grew their bodies with sunlight. All of our petroleum-born plastics, pharmaceuticals, and chemicals also spring from the loins of ancient photosynthesis. In fact, other than rocks and metals, it's hard to find any raw material we use that was not once alive, owing its ultimate existence to plants.

Plants gather our solar energy for us and store it as fuel. To release that energy, we burn the plants or plant products, either internally, inside our cells, or externally, with fire.

For my money, the discovery of fire, as ballyhooed as it was, was vastly overrated. Fire was fine for a while—it kept us warm and cooked our meat. The problem is, we've never gone beyond fire— combustion in furnaces or in engines is still the primary egg in our energy-producing basket, and it hasn't brought us one inch closer to living sustainably. Instead, torching old fuels has led to rising carbon dioxide (CO_2) levels, calving Antarctica icebergs, swelling ocean levels, and the hottest decade on record.

When we burn oil, gasoline, and coal, we release great quantities of carbon that was locked up and compressed during the Cretaceous Period. The giant ferns and dinosaurs of those days decomposed in oxygen-starved conditions and never had a chance to complete their decay cycle. Now we're finishing the job with a bonfire, consuming in a year what took one hundred thousand years of organic growth to form. Like a huge bellows, our bonfire breathes in oxygen and exhales an unearthly quantity of CO_2, a greenhouse gas.

A flux this extreme in a closed system like our biosphere poses the same danger you would face if you burned the furniture inside your house with the windows closed. For the last one hundred years, we've been doing just that—burning the heirlooms made from ancient sunlight, ignoring the fact that contemporary sunlight was streaming in every window. Instead of feeding dead plants to our fires all these years, perhaps we should have been studying the living ones, carefully copying their magic.

AN UMBILICAL CORD TO THE SUN

Though neither popular nor profitable in the shadow of still-spouting oil rigs, the idea of sun-wrought energy has grown tendrils in great minds for many years. Back in 1912, an Italian chemistry professor named Giacomo Ciamician wrote in *Science* magazine about a world in which smokestacks would be felled to make way for forests of clean glass tubes, which would mimic the "guarded secret of plants" and photosynthesize the fuel we needed.

How close have we come to Ciamician's dream? Eighty years later, we have acres of shimmering solar cells made of silicon, a material never found in the blueprints of green plants. After first testing them in the panels of spaceships, we now use photovoltaics (PVs) to pump water, light homes, run laptops, charge batteries, and supplement the electric grid. PVs can cover a rooftop or make digital numbers dance in the tiniest of calculators, but they won't do actual chemistry (making storable fuel from light) the way plants do. And although they're smaller and more affordable than when they first came out, photovoltaics are still nowhere near as compact, efficient, or incredibly cheap as the organic modules assembled by plants. Which brings up another point of envy. Every morning, as our technicians don their white suits and static-free moonboots to assemble high-tech solar cells in toxin-laden factories, the leaves and fronds and blades outside their windows are silently *assembling themselves* by the trillions.

After all these years, and despite the deluge of photochemistry papers published every week across the world, the secret of photosynthesis remains guarded. Fragmentary glimpses of the process reveal themselves, but the working model is still riddled with black boxes (unexplained parts of the process) and mystery molecules code-named Q and Z.

Part of the problem is that the actual harvesting of energized particles of light (photons) is not mechanical in a macroscopic way, a way that we can see with our naked eye. Our strongest electron microscopes can go only so far, showing us *where* photosynthesis occurs, but not how. The "gears" of photosynthesis are molecular, composed of groups of atoms that fly below the radar of even these fantastic scopes. Consider that in the small duckweed that floats atop my pond, there are fifty thousand chloroplasts (the cell-like organelles where photosynthesis occurs) for every square *millimeter* of leaf. Each chloroplast contains a complex network of membranes filled with molecular pigments and proteins all arranged in a fantastically precise choreography. At least, that's what our best guess tells us. For higher plants like duckweed, we're still waiting for actual pictures. In the meantime, we infer the process, build theories, and hunt for proof.

Despite our incomplete knowledge, the spirit of Ciamician still soars in a cadre of artificial photosynthesis researchers. These investigators believe we know enough about the guarded secret to begin building a reasonable facsimile, a solar cell of molecular proportions that will turn light energy into electricity, into a storable fuel, or into the spark we need to do chemistry at room temperature and in water.

Each lab seems to view the guarded secret and the way to mimic it in a slightly different way. Some rally behind the cry of "Charge separation!" Others say, "We need to build an antenna!" Still others shy from using organic building blocks and instead aim to remake nature's design in inorganic form. Each lab is taking a different tack across that great ocean of promise, like boats of different designs in a great America's Cup of science.

In 1990, I was delighted to read that one team in Arizona had pulled ahead. They had actually hitched together an organic molecule modeled on a photosynthetic reaction center, and it had rivaled the quantum yield of photosynthesis! They were rounding the buoy with excited shouts and horn blasts—papers in the prestigious journals *Science* and *Nature*. In March of 1994, I pulled alongside their boat and climbed on.

If you had to dream of a place to pull down the sun, the Arizona State University campus in Tempe would be the perfect setting. Fresh from a Montana winter and still peeling off my parka, I was intoxicated by the sounds of a southwestern campus: the thunk of tennis balls, the laughter from flower-filled grottoes, the incessant birdsong in the palms. I showed up at the Center for Early Events in Photosynthesis smiling like a cruise passenger who's just come updeck for the first time.

But it was hardly a vacation for J. Devens Gust, Jr., and crew. They had just gotten word that the deadline on their major National

Science Foundation grant had been bumped up, and drafts were fly-ing between offices like sideways snow. Despite the pressure, Gust—chemist, professor, and leader of the center—crafted a schedule that would allow me to meet experts from each facet of their work, from the folks who disassemble the real photosynthetic powerhouses to those who assemble the mimics from scratch. As Gust explained, the team held in aggregate what would be too onerous for a single scientist to know, from an understanding of the "quantum uncer-tainty of electron movements in the near-red spectrum of light" all the way to "how the corn plant in Indiana likes its soil, and why." Labs on one floor held glowing jars of some of the world's most ancient bacteria, while in the seismic-steady basement, cutting-edge lasers hummed. On the floors in between, seemingly ordinary or-ganic chemistry labs cooked up molecules that were closer to re-sembling nature's solar collectors than anything else ever made.

My tour at the center was a mental decathlon of sorts, each conversation stretching my understanding of all that is involved in mimicry of this kind. Each team member knew photosynthesis from his or her own scale or discipline or means of measure, but as a whole, they worked as a single organism. And that organism, I got the distinct impression, was in the race of its life.

Biochemist Thomas A. Moore, a leprechaunish baby boomer who tries his best to be a curmudgeon, is frowning at his computer screen when I walk in. Epithets in a soft Texas accent. As if in response, his Macintosh lets out with a guitar riff: "That'll be the day/when you say good-bye-yie-yie/ that'll be the day . . ." On his growl, it backs down.

"It's telling me to get to work," he stage-whispers, "but we'll ignore it."

This seems to please him. Tom Moore is the kind of person who rubs his hands together when he's about to dive into something—a debate, a good meal, a prickly scientific question. There's a certain gusto with which he tears off pieces of life and chews them up. When I ask him to explain photosynthesis, he visibly brightens and (after *how* many years of teaching?) literally leaps up to the white board and starts drawing. "It's amazing," he tells me. "Being able to mimic even a small part of this process reassures me—I say to myself, see?, this isn't magic."

Magic or not, mimicry doesn't diminish the wonder that Moore obviously feels. Every now and then, in between bursts of impassioned sketching—formulas, cells, bacteria, leaves—he says, "I have to go soon." Instead, the clock hands spin, and I learn how the sun turns light into life.

ELECTRON PINBALL

Moore tells me that sunlight is like a drizzle of energy particles, and the job of each green plant, blue-green alga, and photosynthetic bacterium is to capture those particles and put them to work. To help increase their odds, these photon harvesters spread out an array of light-sensitive pigments—chlorophyll a, chlorophyll b, and carotenoids—that act like antennas for the sun's energy. The atoms in each pigment are arranged in the shape of a lollipop—a ring atop a stem. Hundreds of lollipops are embedded in the skin (membrane) of a fluid-filled sac called a thylakoid. Hundreds of thylakoids are stacked up like water balloons inside each chloroplast. Chloroplasts, which make a green plant green, are packed by the hundreds of thousands, if not millions, in even the tiniest leaf.

When sunlight hits these chloroplasts, the lollipop antennas in the skin of the thylakoids grab the packet of energy, then funnel it down to one of the "photosynthetic reaction centers," also embedded in the thylakoid's skin. Each reaction center is a sprawling, ten-thousand-atom assembly with its own set of two hundred lollipop antennas. At its heart is a pair of two highly sensitive pigment molecules that do the actual absorbing. Label this Photosynthesis Central, where light becomes food for life.

Now come in for the close-up. Zooming around this chlorophyll—around all molecules, for that matter—are electrons in orbit, just like the ones you see in those 1950s logos for Atomic cleaning detergent. These electrons are negatively charged particles, the very same ones that, when they cooperate in a flow or current, will toast your English muffins for you. To picture photosynthesis, you have to keep your mind's eye on this moving cloud of electrons. When a leaf absorbs the energy of the sun, some of those electrons zooming around the chlorophyll pair get so excited that they start to migrate to other molecules, setting off a chain reaction in which water is split, oxygen is freed, and carbon dioxide is turned into sugar. In a leaf like the duckweed, it takes two different kinds of photosystems (PS I and PS II) to accomplish this solar alchemy.

Each photosystem stakes out its own portion of the light spectrum. Photosystem II, for instance, absorbs wavelengths that are 680 nanometers long (reddish light), and this absorption causes one of the electrons circling the central chlorophylls to hop to a higher energy orbital, like a pinball being sprung into play. Before it can relax back to its old orbital, discharging its energy as useless heat, an "acceptor" molecule stationed nearby snatches the electron away. But right next door to the acceptor, there's another molecule that's an even better acceptor, and *zap!*, it steals the electron. The electron continues traveling like a hot potato, tossed molecule to molecule away from the chlorophyll. In a few hundred trillionths of a second, a negative charge winds up at one end of a chain of acceptor and donor molecules, and a positive charge winds up at the other. The positive charge is actually a "hole" on the central chlorophyll, created when the electron was whisked away.

Since nature abhors this sort of hole, a nearby molecule code-named Z donates an electron and resets the chlorophyll, sort of like a pinball machine reloading with a new ball. Soon it's off to the races again, with another energetic photon of light captured and a new electron being sprung out of its orbital and into play.

In the meantime, the first hot-potato electron that has been traveling from acceptor to acceptor now jumps the pinball table entirely and goes to the other photosystem, PS I. There it meets a central chlorophyll that has recently absorbed a photon of light (700-nanometer wavelengths) and sprung its own electron into play. That leaves it with a hole, which is conveniently reset by the electron hopping over from PS II. Again, there is a hot-potato toss in PS I as the electron moves from one acceptor molecule to another. The electron eventually moves to the outside of the thylakoid membrane, while the positive charge (all the way back at particle Z in PS II) remains close to the inside of the membrane.

At this point, Moore wheels around and points his marker at me. "And what do you have when you have a positive charge on one side of a membrane and a negative charge on the other?" He's like a demented game show host. I have no idea. "MEMBRANE POTENTIAL!" he shouts, as if we've hit Double Jeopardy.

Every now and then, you discover a scientist's true fetish, the concept that absolutely floors them. Given the chance to explain it to the uninitiated, they stop flatfooted for a moment. There is so much crowding at the door waiting to get out—how will they begin?

"The difference," he goes slowly, patiently, "between a dead bacterium and a live one is membrane potential. In living cells, the concentration of chemicals or charges inside the enclosing membrane is different from the concentration outside. The law of entropy says that all systems want to go to a position of lower energy—they want to equalize uneven gradients or concentrations. That's why a spot of ink breaks up in water—the concentrated ink molecules diffuse into the water and the water molecules diffuse into the ink. Once the concentrations are equal, the system can relax.

"A process like photosynthesis actually creates unequal gradients. It moves negative charges to the outside of the thylakoid membrane, leaving a buildup of positively charged ions inside. This polarizes the membrane, making the inside of the sac different from the outside. The charges on either side of the membrane want to recombine, to release their energy and relax; that would be a downhill reaction, the most natural thing in the world. But because the membrane is in the way, the tension remains high. Your car battery does the same thing—it separates charges as a way of storing energy. Living cells, like cars, can use that energy potential. They use it to import nutrients, to get neurons to spark, to get cells to talk to one another, or to get muscles to move. On a cellular level, life lives in the tension between unequal concentrations, unequal charges. Membrane potential equals chemical and electrical potential equals life."

At this point, having not cracked a cellular biology textbook for many years, I felt the concept wobble out of my reach a little. Moore, the consummate teacher, returned to the leaf.

Membrane potential has a lot to accomplish in plants, namely the feeding and fueling of an entire planet. First, there's the splitting of water. With each electron that the PS II chlorophyll springs into play, the molecule Z donates one of its electrons to "reset" chlorophyll. Z eventually donates four electrons to PS II. To reset its positive holes, it teams up with a water-splitting complex that strips four electrons from water (H_2O). This liberates oxygen, which percolates out of the leaf, and hydrogen ions (H^+), which get stuck inside the thylakoid sac. Hydrogen ions, being positively charged, want desperately to even the score and get to the outside where negative charges reside.

In the meantime, at the outside of the membrane, one shuttled electron after another is handed off to a molecule called NADP$^+$ (nicotinamide adenine dinucleotide phosphate). This hand-off transforms NADP$^+$ into the electron carrier NADPH, which has mighty "reducing" powers (the ability to give electrons to other compounds). This means that in the next stage of photosynthesis, the so-called dark phase, NADPH can give electrons to CO_2, and thus "reduce" it to sugar, CH_2O. But it can't do that without a sidekick—a molecule that will provide energy.

"And here," says Moore, "is where the membrane potential comes in."

The only way for the trapped hydrogen ions to get out of the thylakoid sac is through an enzyme "channel" called a coupling factor. In textbook cartoons, it looks like a toadstool, with a stem spanning the membrane and a bulbous head sticking outside. As the plus charges escape through this coupling factor, they extract a toll—they turn a compound called adenosine diphosphate (ADP) into adenosine triphosphate (ATP) by adding a third phosphate. This third phosphate is hitched to the other two with a high-energy bond, and it is here that the energy of the sun is stored. During the dark reactions, ATP's high-energy bond is severed and the energy is used to turn CO_2 into sugar.

The chemistry that stored this energy couldn't have occurred without two charges, a plus and a minus, being banished to opposite ends of a membrane, sent packing by the power of ordinary, garden-variety sunlight. Anytime you have a positive and negative charge separated like that, you essentially have a battery, a battery powered by the sun.

Moore takes another deep breath. "We began to wonder if we could make a solar battery by hooking a sun-sensitive pigment to a string of donor and acceptor molecules. We wanted two things. First, we wanted to get charge separation—a plus at one end of the string and a minus at the other—and second, we wanted the charges to stay separated long enough for us to accomplish work."

"Work" could take many forms: 1) Hook wires to the ends of the molecular string to get an electric current, 2) use it to split water and produce clean-burning hydrogen gas, 3) use it as a power pack to drive solar-based manufacturing, or even 4) use it as a switch for computing near the speed of light.

"One day, we may even convince our string of molecules to go into the membrane of an artificial cell," says Moore. "Instead of boiling chemicals for several hours in toxic solutions to make plastics or other products, you could build a tiny reaction vessel, give it a power pack, and stand back so you're not blocking its sun." What's science fiction for us—clean-burning fuel and chemistry in sunlight—

is commonplace for plants. Somebody needs to tell Aristotle that the gods are in the kitchen after all.

SOLAR ALCHEMY

Speculation is a lovely sport, but as any of the scientists at the center will tell you, it's one thing to work out a prototype of a donor-pigment-acceptor device on paper. It's quite another animal to actually hook the molecules together so they'll transfer electrons. Putting theory into practice means taking small steps into uncharted (by humans at least) terrain, working from maps that are sketchy at best. But considering the fact that photosynthesis produces 300 billion tons of sugar a year, it is undoubtedly the world's most massive chemical operation. Every pine needle and palm leaf can do it. The more I thought about it, the more amazed I was that no one had taken Ciamician up on his dare. How hard could it be to duplicate the first few picoseconds, the electron transfer part? And why haven't we done this before?

That was before I saw the molecular map of a photosynthetic reaction center. Devens Gust has a full-color reproduction of a purple bacterium's reaction center in his office, and he and I spent a while just admiring it. The visual was relatively new to those who had been studying photosynthesis for years. As Gust examines it, his black eyes focus like a hawk's on a gopher hole, and for a moment, I lose him.

Devens Gust is a deep river of a man, possessing a trademark calm that plays well in combination with Tom Moore's quick passion. While Tom and his research partner Ana (who is also his wife) are usually in the office long after the dinner hour, Gust closes his door at five and rarely shows up at the lab on weekends. "Devens can get more done in a forty-hour week than most of us get done in seventy," Tom Moore tells me. Before I left, Gust pulled out a map and helped me plan a road trip across Arizona, showing me where I could find Anasazi ruins that the tourists pass by. "That sounds like Devens," said Moore. "He actually has time to go hiking!" And time enough, with a deadline looming, to show me the heart of what inspires his team.

The reaction center is a startlingly beautiful device, composed of several chemical groups called cofactors, set like jewels in a tangled bird's nest of protein—what scientists call the protein pocket. When you connect the cofactor dots, you get something that looks like a wishbone, with a chlorophyll pair in the center and two curving bones of cofactors facing one another with near mirror symmetry. Ten thousand atoms are choreographed in the membrane just so, with a geometry that allows them to play the pinball game of electron transfer. Faced with a blueprint this complex, what steps would one take to build a solar battery from scratch?

"We knew it would be ludicrous to try to duplicate anything as complex and finely evolved as this," says Gust. "Nature has a three-billion-year jump on us here." The purple bacterium we are admiring is a sun-harvesting microbe that researchers routinely study for clues to photosynthesis. It's sort of the fruit fly or *E. coli* of photosynthesis research because it's easy to culture, easy to read genetically, and structurally simpler than green plants. It's more akin, they believe, to the first photosynthesizers that arose three billion years ago. Instead of two photosystems, purple bacteria make do with one that is analogous to PS II. "Because people have been studying the purple bacterium so intensively, its reaction center has fewer black boxes than any other system; it's the closest thing to a blueprint that our team has.

"Our goal was to pare down the reaction center and model only its essence. We wanted our device to *work* like this, even though we knew it would not *look* at all like this." The natural reaction center, for instance, uses that tangled scaffolding of proteins to embed and hold independent cofactors. Not wanting to tackle anything as complex as a protein pocket, the team took a different route. In their device, the cofactors float in a beaker of liquid, bonded to one another with great care via organic chemistry techniques.

"The bonds must duplicate the scaffolding's magic—they have to hold the molecules at the correct geometry and distance from one another to provide the proper pathways for electron transfer. To accomplish this feat of mimicry," says Gust, "we peered over nature's shoulder, tried something, peered over nature's shoulder again. Lately, we've been going to Neal a lot."

Neal Woodbury is a chemist turned photosynthesis sleuth who uses genetic scissors and glue, laser scopes, and millions of bacteria to do his detective work. "Have you seen them?" asks Woodbury, leading me across the hall to the bacteria growth lab. The lab looks like any college biology lab, with long benches and overhead shelves crammed with Bunsen burners and glassware. He squats, reaches below one of the lab benches, and opens the double doors, revealing a series of warm, brightly lit chambers filled with large jars.

They remind me of the jars you find in country saloons, bobbing with pickled eggs. Some of the jars contain a sludgy brown substance, while others are a mossy green. He moves these aside to find a jar of purple bacterium, *Rhodopseudomonas viridis*, the color of an Easter egg dye, but thicker. As he holds it up to the light, I see no movement, no sculling backstrokes or twirling whipcords. These bacteria, I remember, are far below my capacity to see, and the jar that Woodbury is holding must contain billions of individuals. As we speak, reproduction and attrition swell and shrink the population.

For a long time, Woodbury tells me, they had to work from inference, guessing how the cofactors were positioned, because there was no molecular picture of the reaction center. "One of the most dramatic advancements in photosynthesis in this century was to finally get our pictures back from developing and see a bacterial reaction center molecule by molecule. The reason it took so long is because this assembly we are dealing with is so tiny—taking its picture with something as big as a light ray would be like bouncing a tennis ball against a poppyseed."

Instead, scientists had to use small X rays to take the pictures. The technique is called X-ray crystallography, because the molecule to be "shot" is first crystallized—its molecules are lined up so they are all facing the same way, in dress-parade perfection. The X-ray beam passes through the molecule, and the pattern of the diffracted X rays is recorded as an array of spots on a photographic plate. This pattern tells scientists how the atoms are arranged in the molecule—what's next to what. The toughest part of the process is getting the molecule to crystallize—a protein crystallographer can easily spend eight to fifteen years trying to get a good crystal and picture of one type of molecule.

The key to getting a good crystal is to completely dissolve the molecules in water first. With proteins that live in membranes, this is no mean trick. Having an affinity for fat (membranes are double layers of fat) but not water, membrane proteins simply clump up in the bottom of a beaker instead of dissolving. It wasn't until scientists learned to hook water-loving helper molecules to them that the reaction centers were able to blossom in water and finally have their picture taken.

The scientists who achieved this feat (German chemists Hartmut Michel, Johann Deisenhofer, and Robert Huber) won the Nobel Prize in Chemistry for it in 1988. "Until then," says Woodbury, "we had been guessing about what elements were in the reaction center, and how they might be oriented in relation to one another. The pictures showed us exactly how nature's geometry works to enhance the transfer of electrons. Now we have some definite plans to inspire us."

Still, when Woodbury looks at these bacteria, the picture in his mind's eye is much more detailed than what the rest of the world sees from the new molecular maps. Even before the maps were drawn, he and other geneticists had been probing purple bacteria with their own set of tools, sequencing the proteins and making deductions based on carefully controlled mutations. "I know every amino acid in that protein pocket," says Woodbury. "But knowing what they are and knowing what they *do* are two different things. These days, we want to go beyond mere structure. We want to know how structure affects function—what exactly makes it work so well. I find this out by distorting or even 'turning off' one piece of the structure at a time, through a process called mutagenesis." Simply put, Woodbury uses biotechnology to create mutant bacteria with a specific defect in their reaction centers. "The question we ask is, how does this specific change affect their ability to photosynthesize? That's how we learn which parts of the reaction center are most important."

The purple bacterium cooperates in a notable way, which is what makes it such a model organism to work with. Not only is it a simple system, consisting of just one kind of reaction center, but it is also ambidextrous when it comes to garnishing energy from its world. It can photosynthesize one moment, then switch to oxidizing its food through respiration just like we do. As Woodbury says, "That flexibility means we can tinker with its photosynthetic mechanism, and even impair it a bit, without running the risk of killing the patient."

I try to imagine the size of its reaction center, given the fact that it is tucked into a membrane of a bacterium that Woodbury tells me is only one to three microns long. Several thousand of these bacteria could fit into the period at the end of this sentence. Now imagine that the reaction center inside this bacterium is itself only about thirty angstroms by eighty angstroms. An angstrom is one tenth of a billionth of a meter. If you had a string of beads, each one an angstrom, and you wanted to span an inch, you would need to string together 250 million beads. Now tick off the first thirty of those beads, and you have an idea how wide the reaction center is. Tick off eighty, and you've got the reaction center end to end.

"An electron moves down one side of that wishbone-shaped reaction center at a speed that is equally astonishing," says Woodbury. "It's measured in picoseconds—trillionths of a second." To comprehend this tiny number, consider that one picosecond is 1×10^{-12} seconds and the age of the Earth is about 1×10^{12} days. That means a picosecond is to a second as a day is to the age of the Earth. And it takes only a few hundred of these picoseconds for an electron to make it from the inside of the membrane to the outside. By the time you can form a thought, charge separation could have happened many millions of times. How do you spy on a molecular complex that small, and capture a process that fast in the act?

The answer to the size question is that you don't spy on one molecular complex; you spy on a whole test tube of reaction centers at once. The trick is to give them a "start" pulse of light so they all begin photosynthesizing at the same time. That way, what's happening at any one moment to the group is also what's occurring in each reaction center.

You tackle the timing quandary with ultrafast laser pulses that flash on to take the reaction center's "picture" at various stages of electron transfer. To see this ultrafast photography for myself, I went down to the laser room where Woodbury had laid out a track for the laser light to go around. Imagine a Christmas train set with light moving around it instead of trains. There were beam splitters, mirrors, and laser beams of various colors focused on vials of purified photosynthetic reaction centers. The starting gun is a burst of coherent light (oscillating up and down in perfect step, at the same phase and wavelength). It excites the reaction centers and causes them to begin their handoff of electrons. While that happens, Woodbury probes the vial with a second beam to see what's up.

"At rest, every molecule will absorb at a precise wavelength and then fluoresce—emit light as it releases the energy. But when that molecule is excited by sunlight, for instance, it will change its shape and absorb and fluoresce at a different wavelength. [This is the idea behind mood rings—when the chemicals heat up and change shape, they absorb (and reflect) a different color of light.] This 'spectral signature' changes continuously as a molecule moves and participates in photosynthesis, the electrons hopping from one spot to the next. By tracking these changes in spectral signature, we can spy on what the molecule is doing."

After pulsing the vial of reaction centers with a "start" light, Woodbury dials a new wavelength on the laser and begins probing with that beam. He snaps picophotos at discrete intervals of time, looking for fluorescence. "The molecule changes shape as it moves through the reaction. We're watching carefully, and when the molecule absorbs the probe beam and emits light, we note the time. This tells us that at one point three picoseconds, it had a spectrum like this, and was in this particular stage of the reaction. We then repeat this probing with different wavelengths of light to get a complete picture, really more of a movie, of how the molecule changes through time. The reaction centers from our mutated bacteria will function differently from wild bacteria. By comparing the movie of the mutant's changes with the movie of a wild reaction center, we try to guess how the mutation has affected photosynthesis."

Woodbury induces the changes in the wild reaction center by rewriting the genetic blueprint (editing the DNA sequence). "When I pull out a piece that makes photosynthesis shut down entirely, I figure, here's something important, and I go tell Devens and Tom." Devens Gust puts it this way. "It's as if Neal is digging inside a computer and removing random parts of software programs. Say we want to know what makes word processing tick. One day he removes the fonts, and we can't type anymore. So we say, fonts must be important. Let's go model them."

No one wakes up one morning and decides to model something as big as a reaction center; the quest grows organically from much humbler beginnings. Years ago, Tom and Ana Moore were hot on the trail of antenna function—the satellite dish that expands the plant's reach. Tom had done his graduate work on carotenoids (the pigments in antennas), which at that time had not been all that well characterized or mapped. He and Ana were trying to isolate carotenoids from living systems to see how they worked, but it was proving difficult. Devens Gust, in the meantime, was working with molecules called porphyrins, which are cousins of chlorophyll and also appear in antennas.

At lunch one day, Gust and Tom Moore, who had never worked together, began talking about their separate but mutual antenna problems. They thought, why don't we try to hook a carotene to a porphyrin and build a simplified antenna? At that time, there were no pictures of how these components were oriented in relation to one another in real life, so Moore and Gust made educated guesses and tried to chemically bond these molecules to test their hypothesis. "It was a grueling task, and grad student after grad student gave up in frustration. Finally we hired a Ph.D. student from the University of Montana, Gary Dirks, who locked on this thing like a pit bull. He lived in the lab until he finally found a way to put them together at just the right orientation, and they worked!" They had guessed that the carotene and porphyrin would have to have their orbitals overlapping somewhat to help the energy resonate from one to the other, antenna style. Although their idea made sense to them, it went against conventional wisdom at the time. "When the pictures of purple bacteria came out, we were thrilled to see that the antenna elements in our artificial device were indeed oriented at almost exactly the same angle and distance as the real antennas," says Moore with pleasure. "We were spot on."

The Dyads

But *energy* transfer wasn't *electron* transfer; having grabbed one brass ring made them want to try for the other. Already, other boats had pulled ahead in that leg of the race. Paul Laoch at Northwestern and a group in Japan had managed to build a dyad (a two-part molecule) that would transfer an electron from an excited porphyrin to an acceptor called quinone. Instead of relaxing back to its old orbital around porphyrin, the excited electron suddenly had a "competing pathway"—a better offer—in the form of the orbital around quinone. This second orbital was especially inviting because it was close at hand and slightly lower in energy, like a basin in the energy landscape. The trick was to bond a donor and acceptor so that their electron orbitals overlapped.

"It's as if the dyad builders had dug a riverbed between the two molecules." Unfortunately, the "slant" of the riverbed was not great enough, and it allowed the electron to flow both ways. After a brief separation of plus and minus, the electron soon found its way back and the charges recombined in a burst of heat, wasting the energy before it could be used. "They had a pretty good yield [the percentage of photons that successfully triggered charge separation], but the charge-separated state lived only ephemerally—one to ten pi-coseconds." Since that's too short to get chemical work done, it wasn't a good mimic of photosynthesis.

"Our task was to get the charges to separate quickly and then hold them like that—to slow down recombination. Putting some physical distance between the plus and minus charges seemed like a good delay tactic. We asked ourselves, what if we were to add an-other molecule to the donor-acceptor dyad and make it a donor-donor-acceptor string, a triad?" Gust and Moore had already had luck hooking carotene and porphyrin together in a donor-donor pair. By adding a quinone as an acceptor, they would create a triad. In 1979, they set their sails.

The Triad

On paper it looked like a straight shot to the buoy. But in the lab, winds are fickle, and nothing is as smooth as you might imagine, especially when the waters are uncharted. Dr. Ana Moore would be the actual builder of the molecule, the wet-lab chief who would put it together one agonizing organic reaction at a time.

If Gust has the eyes of a perched hawk, then Ana Moore's are a raven's—cocked, curious, and piercing. Like Tom, she is riveted by her work and tells me that she dreams solutions to knotty prob-lems in her sleep, or they occur to her unbidden in the shower. For our talk, we go out to one of the terra-cotta benches that I spied on the way in. Here we are bathed in unadulterated Arizona sunlight, contemplating a process that might make the power of that sunlight as available to us as it is to the nearby vines. Ana Moore illustrates the future as she sees it, filling my notebook with chemical graffiti.

More than anyone else, Moore speaks with an engineer's sensi-bility, as if the chemical groups, which were so hard for me to vi-sualize, are actual bone and brick, strung together with hinges and joints. She talks me through the synthesis process in a strong Argen-tinean accent, speaking faster and faster as her excitement builds, like a roller coaster after the crest.

"We decided to bind the groups together with amide bonds, which is what amino acids are joined with. Amide bonds are stable and versatile—we figured these bonds would keep our molecule strung out in a line, rigid enough so that it would not fold back on itself and mechanically recombine the charges. The only problem is that forming these kinds of bridges takes many, many steps."

Organic synthesis is an art, she says, like gourmet cooking. Acquiring a "good touch," a feel for how and when to time and sequence your reactions, is not something that can be taught. You must simply do it for years and years. A good synthetic organic chemist is grown, and a team like the ASU group would be stuck in the irons of theory if it were not for a good synthesist.

Mention Moore's name to others in the ASU team, and you'll hear the words *wizard, magician,* and *miracle worker.* I repeat this to her and she laughs, not understanding the fuss. "You know why I do this?" she asks me. "I love to build molecules. Once I know a compound exists in nature, I have to build it just to see if I can. But to build this one that does something, that's even better."

Building a molecule synthetically means keeping track of many different reactions happening in beakers throughout the lab over many months. To prepare a chemical group for bonding, you must first give it a chemical handle that other chemical groups can home in on and bond with. While that's happening, you're also adding a handle to the next group you want to include. To make sure you get only the reaction you want, you must protect certain sites on the chemical groups, in a kind of masking procedure. Once everything is masked and equipped with handles, you forge your first bond. If that's successful, you go in and deprotect the masked sites and begin all over again—adding a handle, protecting, bonding, and deprotecting. Each of these steps requires a special reactant—a chemical bath bubbling at just the right temperature for just the right amount of time. It may involve dozens of stages, adding layer upon layer until you have built your molecule.

"If one step bombs," she says, "you have to clean out your beakers and start all over again.

"Right in the middle of building the triad Tom went to Paris for a sabbatical, and I got a position with the French Museum of Natural History. Devens was also working over there, along with a colleague of ours named Paul Mathis. They were all working at Saclay, a nuclear facility that happened to be the site of the largest photosynthesis research lab in Paris." The nuclear connection to photosynthesis is not as incongruous as it sounds. Most of what is known about photosynthesis was learned by putting radioactive tracers on CO_2 and then following the carbon through to its products in the leaf.

Moore took her plans for the molecule to Paris and began building it in a bare-bones wet lab at the museum. "Every night I would bring the work-in-progress home with me and put it in the fridge. I had two small children in day care, and by the time I had gotten to the office, I had been on five subways, all with my precious vial. It took me a year and a half to finally assemble a molecule that we thought would work. The museum didn't have the spectroscopes we needed for testing so I had to send the vial out to Saclay for Tom and Devens to test."

Tom Moore picks up the story. "We knew that when we irradiated it, if it was working properly, the negative charge would run one way, leaving a positive charge at the other end. The positive charge on one end would cause the assembly to absorb light at a particular wavelength, so that when we had that final product (a charge-separated state) we'd see a large jump on our detector instruments. Sure enough, when we probed it with a certain wavelength, we saw an enormously large signal. We were jumping up and down, and about to call Ana, when the technician came out red-faced and told us he had set the probe at the wrong wavelength so that what we had just seen was not accurate. It was a tremendous letdown and we thought, that's it. It's not going to work. But then we set the thing up correctly and, lo and behold, it worked. The signal was even stronger! Ana took the train out to Saclay, and, being the Doubting Thomas that she is, she wanted to run the control again."

"When I saw it with my own eyes," says Ana Moore, "I knew it was true."

Perhaps the most amazing thing was how long they were able to keep the plus and minus separated in the triad. "Before this, the longest charge separation (using a dyad) had lived ten to one hundred picoseconds before collapsing together in a burst of heat. There was no way to grab hold of the potential. But with the triad, we watched the clock tick off the digits and we couldn't believe our timing. It lasted and lasted—two hundred to three hundred nanoseconds, ten thousand to one hundred thousand times longer than the dyad! For the first time, we had sufficient distance and staying power to envision doing some real chemistry at the ends."

Standing in Saclay, France, thousands of miles from home, the photosynthesis mimics stared at the instruments, then at one another, then back at the instruments. They'd done it! They'd glided round the buoy in the race the whole world of photochemistry was watching. Gust presented the paper at a Gordon Research Conference in 1983, setting the pace for other boats of researchers for the next decade.

The Pentad

After that, triads of all shapes and descriptions were developed. Everyone was trying to better Gust's and the Moores' separation time and yield. For their part, the trio was already thinking of ways to go beyond the triad and separate the charges even farther. Ana Moore relates, "When we came back home, we hit the ground running. We built it to a four-part molecule, and finally, we went to the pentad, a five-part donor-donor-donor-acceptor-acceptor molecule. That's our best effort yet. With the pentad, we achieve a quantum yield of eighty-three percent, meaning that for every one hundred photons we pump into the system, eighty-three of them cause a charge separation. Photosynthesis runs at ninety-five percent, so we are creeping closer. Best of all, the charge-separated state of the pentad lasts even longer than the triad. We're tweaking to improve it all the time."

The pentad's chemical signature is written like this: $C-P_{zn}-P-Q-Q$. On the far left is a carotene, then a porphyrin molecule with zinc, then plain porphyrin, then a naphthoquinone and a benzoquinone. The donor-acceptor lineup looks like this: D-D-D-A-A. Each of these molecules has a unique shape and electronic "personality" and, therefore, a unique affinity for accepting or giving away electrons. To make sure the slant is in the right direction, so the electron doesn't find its way back too soon, the left-to-right lineup features better and better acceptors, each with a lower profile in the energy landscape. The carotene at the far left is the best donor; it's highest in the energy landscape and most eager to give away its electron. The quinone on the far right is the low spot in the energy landscape and therefore the best acceptor. The electron goes from one to the other like a ball bouncing downstairs, ultimately coming to rest on the last quinone.

In the pentad, the artificial photosynthesizers have added another dimension besides pure electron transfer. There's a tiny mimic of the leaf's antenna in the P_{zn}-P pair. When light strikes the P_{zn}, *energy*, rather than an electron, heads over to the P. P then reacts to this energy by donating an excited electron to the first quinone, which in turn transfers it to the second quinone. Each positive charge or "hole" that is left is neutralized or "filled" by an electron from the molecule to the left.

To anthropomorphize like crazy, let's say the five chemical groups (5-4-3-2-1) are spectators at an outdoor concert. They are all wearing lap robes. A breeze lifts the lap robe off of 3. Number 2 is eager for another lap robe, so he steals it away, winding up with two lap robes. Number 1 is even more eager and steals the extra lap robe from 2. In the meantime, poor number 3 has lost his robe and is cold. A generous soul at his left, number 4, donates his robe. Number 5, being even more generous, gives 4 his robe. Now number 5 is robeless (positive charge) and number one has an extra robe (negative charge).

In the parlance of chemical graffiti, the lap-robe shuffle would look like this, with light energy moving left to right, and electrons moving left to right to neutralize holes left behind by donated electrons:

(*)= excitation of energy, (−)= extra electron, and (+)= hole left behind by a donated electron.

Step 1. Light excites P_{zn}		$C\ P_{zn}{}^*\ P\ Q\ Q$
Step 2. Energy transfers from P_{zn} to P		$C\ P_{zn}\ P^*\ Q\ Q$
Step 3. An electron transfers from P to Q		$C\ P_{zn}\ P^+\ Q^-Q$
Step 4. An electron transfers from P_{zn} to P		$C\ P_{zn}{}^+\ P\ Q^-Q$
Step 5. An electron transfers from C to P and Q to Q		$C^+\ P_{zn}\ P\ Q\ Q^-$

Because the shape of a molecule and its interactions with neighbors determine just how likely it is to donate or accept an electron, pentad builders have a variety of "knobs" they can tweak to increase the rate of electron transfer. They can change the chemical structure of the molecules, their distances from one another, or even their interactions with the surrounding medium, which at this point is a liquid solution. Someday, Neal Woodbury speculates, they may even be able to embed the pentad in a membrane, surrounded by protein scaffolding that will further speed or slow down the transport of the electron.

The trick to pentad tweaking is to use a light touch, says Gust. "You don't want to have the energy differences between steps too great, because with each step a little of the initial sun-energy that came into the system is lost. Too large a drop would mean the loss of too much energy. Instead, you want a shallow series of steps, each one dropping down only a little in the energy landscape. Say you begin with two volts from the sun. In our best efforts, we were able to keep a full fifty percent of the energy from each photon we put in. Two volts in, and at the end of the sequence, there is still one volt left to do work. That's right up there with photosynthesis."

By building the pentad, the ASU team proved an important principle. They showed that if you can get charges to travel far enough apart in space via steps that prove more seductive than the natural urge to recombine, those charges will stay separated for a long time. They further showed that if you make those steps shallow enough, you'll buy yourself energy as well as time. The question is, energy and time to do what?

SPANNING A MEMBRANE: CATALYST WITH A POWER PACK

None of the scientists was too anxious to talk about applications. After their last piece of publicity (an article in *Discover* magazine), they began to get calls from people wondering *when* they would be able to buy molecular batteries at Wal-Mart. The team at ASU is quick to emphasize that their research is squarely in the basic realm, and they're happy to leave the actual application work to engineers. "At the center, we are much more interested in perfecting our understanding of nature's mechanisms than we are in building devices," says Gust for the record.

Yes, I say, but if someone, someone else of course, were to build something, what might it be? Disclaimers out of the way, we begin to blue-sky.

None of them thinks that rooftop photovoltaics will be made out of pentads any day soon. In their current form, high temperatures may wilt them and cold temperatures may freeze them, so they're unlikely to last twenty years on your roof. What about lichens, I ask, which contain algae able to photosynthesize at subzero temperatures, or desert plants able to survive quite nicely in the hellish temperatures of Death Valley? The difference, Gust reminds me, is that living plants can replace spare parts when they wear out. No matter how similar its function, a pentad can't do that. Besides, if trends continue, he says, silicon photovoltaics are likely to keep coming down in price to the point where it might actually be economical to have them on your roof.

What sets the pentad apart from silicon cells, however, is its size—at eighty angstroms, a pentad is a very tiny double-A battery that is activated by light. In a world where machinery is fast approaching the molecular scale, there will be plenty of call for vanishingly small batteries. If you could find a way to hook them to a grid, I suggest, you could pour billions of pentads into a can of paint and layer your house with sun harvesters! Or paint the highway system with them! "Try doing that with a rooftop photovoltaic," laughs Moore. Then he arches an eyebrow, looks both ways, and leans toward me. "Do you know what will be really amazing? When we find a way to get this thing to embed in an artificial membrane. *Then* we'll be cooking.

"What we have now is essentially an electron transfer device," he explains. "What we want to do next is what photosynthesis does next, which is to convert charge separation into membrane potential [he never misses a chance to bring this up]. To do this, we have to design an artificial cell, put the molecule in the membrane, and shine light on it. If we can do that, we will have converted light into a voltage across a membrane. Then we can make use of any of the biological paradigms for using potential. Pumping ions, making ATP (the gasoline of life), importing sugars—anything biochemistry does with potential, we can do once we learn to incorporate molecules into membranes."

Scientists already know how to make an artificial cell—they put lipids (the molecules that make up cell membranes) in water and shake them up so that they self-assemble into watery spheres called liposomes. If Gust and the Moores could install their molecule in the skin of one of those bubbles, along with the toadstool-shaped coupling factor that makes ATP, they could shine light on it and make the fuel of life. "Just think," says Tom Moore. "We would demonstrate the production of ATP in a light-driven system."

What to do with it? Moore sighs. "Well, first, I'd stand back and admire it for a long time. Then I suppose we could mimic an uphill reaction that needs energy—like the assembly of a protein. Put in everything a cell needs to make proteins—a ribosome system, DNA, amino acids—and then shine a light on it and see if it will crank out a piece of protein, like insulin." Right now insulin is made by genetically engineering *E. coli* bacteria. The day may come when we could dispense with the bacteria that have to be fed and kept at certain temperatures and instead have tiny nonliving factories—sacs with power packs in their skin. Being fearful of genetic tinkering, even when it's done with *E. coli*, I like this alternative.

Ana Moore, the engineer, thinks of logistics. "Right now our pentad is too long to fit lengthwise across a membrane—the membranes have room for something thirty angstroms long and the pentad is eighty. For membranes, our best shot may be the shorter triads, but first we'll have to improve yield and charge-separation times. Then we have to get the molecule to recognize the membrane, enter it, and line up in the right direction. Of course, we'll have to deal with the interfacial relations between the triad and the proteins it will encounter in the membrane layer—right now it's just floating in solution." As she speaks, I see her mind whirring, plugging words into the grant application.

"Come back next year," Tom Moore teases. "We'll show you how it works."

Neal Woodbury likes to imagine how chemistry might change even without a membrane, if we could find a way to hook pentads to catalysts, those workhorse proteins that float around inside cells, joining molecules together and splitting them apart. Like spot welders, catalysts work with amazing specificity, honed over eons of evolution.

Biochemists have a whole arsenal of nature's catalysts that they can take off the shelf, compounds like DNA polymerase that zips along DNA, making thousands of copies. These biochemical reactions are for the most part thermodynamically downhill. You just mix in the catalyst, and the reaction proceeds without enormous inputs of energy. Biochemistry is like that.

Unfortunately, a lot of the chemicals and pharmaceuticals we manufacture are *uphill* reactions, which we have to coerce with strong chemical baths, high heat, and extreme pressures. What if instead of bulk chemistry consisting of forty or fifty steps, you could go to the shelf and pick out a designer spot welder (a catalyst) with its own power pack (a pentad)? You could mix it with precursors A and B and hit it with light, and it would do uphill reactions, forming AB for you with the kind of specificity that nature achieves. In this way, we would be able to build chemicals efficiently and cleanly, in water, using sunlight as the energy source and producing no noxious byproducts. Now *there's* something we could stand back and admire for a while.

HYDROGEN DREAMS

Finally, if we are to mimic a green plant's real planetary coup, we must find a way to use the light of the sun to run a chemical reaction that would net us a storable, high-energy fuel. With all due respect to plants, sugar and starch are not what we humans had in mind (plants already do a fine job of making those for us). What does interest us is the possibility of producing hydrogen gas from sunlight and water.

Hydrogen is the world's cleanest storable fuel—it can be derived from water, and when you burn it, you release pure water again. Hydrogen is also the fuel of choice in fuel-cell technology. Fuel cells are portable devices that take hydrogen gas and use it to generate electricity, right in your car, for instance. At this point, fuel-cell technology is still an elusive goal—no one can get the chemical reaction to work for more than a few hours. If and when the barriers are overcome, the demand for hydrogen gas will be immense.

The alchemy needed to "crack" water and extract hydrogen gas does not look difficult on paper. Nature does it all the time with the help of an enzyme called hydrogenase. Hydrogenase takes hydrogen ions (H^+) and, with the addition of electrons, makes H_2 gas, which can be bubbled out of the solution. Photosynthesis produces all the needed ingredients. It releases hydrogen ions from water and shuttles electrons into the hands of $NADP^+$, which becomes the electron carrier NADPH. As long as we have hydrogen ions and this constant source of electrons, we should be able to add hydrogenase and collect our H_2 gas for free, right? Unfortunately, it's not that simple. Hydrogenase is not comfortable in the presence of oxygen, and after a few hours of pumping out a product, it is overcome by oxygen, and the reaction grinds to a halt. Technology watchers predict that it's just a matter of time, however, before someone perfects the side reactions. When they do, the world will come looking for a sun-harvesting power pack to provide the charge separation. Chances are the pentad, or an even newer and improved model based on the reaction center, will be on the short list of candidates.

COMPUTING AT THE SPEED OF LIGHT

In the meantime, the most likely application on the horizon is a marriage that's hard to picture: technology mimicked from the world's most ancient organisms breathing life into a brand-new generation of computers. These organic-silicon hybrids, sporting switches the size of a molecule, will make Pentium PCs seem as plodding as the vacuum-tubed ENIAC from the fifties.

Today's computers use a series of switches to store and transmit electronic bits—the zeros and ones of digital code. The switches act like those in a railroad yard. They open to let trains of electrons pass through whenever they receive the right signals. Conversely, some switches can be shut down to block the flow of electrons. What most of us don't realize is how slow and labored this process really is— with a linear series of switches, the computer can do only one calculation at a time, in sequence. Computers of the future will be more like brains—they will have three-dimensional webs of switches. The signals, instead of traveling via electron flow, will be encoded on light waves traveling at, well, the speed of light. Say you want to send the *Encyclopaedia Britannica*—all thirty or so volumes—from Boston to Baltimore. If you send it on today's copper wires and squeeze it into your computer's 28.8-baud modem, it would tie up your phone lines for half the day. That same transmission sent via light waves in a hair-thin optical fiber would show up in less than one second.

To equip these optical wunderkinds, technologists will need light-sensitive switches, the smaller the better. A device like the pentad, which changes its charge distribution (how its electrons and holes are positioned) in response to a certain frequency of light, makes an ideal switch. Hit it with light, and the negative and positive charges will zip to opposite ends of the pentad. When the switch is in this charge-separated state, it actually *changes shape* and therefore absorbs light from a different part of the light spectrum (the mood-ring phenomenon). This means that the pentad can be controlled—it can be flipped back and forth from a state in which it absorbs only red light, for instance, to a state in which it absorbs only green light. In computer lingo, those states are called off and on, zero and one.

Gust and Tom Moore have been daydreaming publicly about the possibility of installing pentad switches by the millions in a durable material. Their articles in computing journals outline the specifics of molecular "OR" gates and "NAND" gates. Here's one scenario: In its charge-separated state ($C^+ P_{zn} P Q Q^-$), the pentad will absorb light at wavelengths measuring 960 nanometers (nm). In a switch where light is hurtling through at 960 nm, a charge-separated pentad would block that light by absorbing it and stopping its transmission. Essentially, it would switch off the flow. Conversely, a pentad in its relaxed state wouldn't absorb the 960 nm light and would therefore let it pass. We could toggle these molecular switches from a relaxed state to a charge-separated state by hitting them with a preparatory pulse of light, essentially opening or closing the gates to the transmission of bits and bytes.

This last application may seem far from the inspiration of photosynthesis, until you remember that by finding a new application for the machinery of photosynthesis, we are being the ultimate biomimics. "Nature is famous for retrofitting an existing technology to accomplish many different things," Tom Moore reminds me. With a few modifications, he says, the same mechanism that turns carbon dioxide plus water plus energy into sugar and oxygen is simply run in reverse whenever we eat a salad or a stroganoff. We take sugar and oxygen and break them down to energy, carbon dioxide, and water. What these mirror reactions have in common is what cuts across the plant and animal kingdoms: the miracle of membrane polarization. In fact (I'm sounding more like Tom Moore every minute), it's a common theme in all biological functions, including thinking. As you read this sentence, the membrane potential in your nerve cells is helping you send signals, process information, and, in short, compute. Suddenly, playing a game of Tetris on an artificial photosynthesis computer doesn't seem so strange—it's just another acorn that has fallen, and not all that far from the tree.

Later that week, thinking about all this as I make my way among the Anasazi ruins that Devens Gust had turned me on to, I begin to smile. After all these years, we are only now looking to leaves as a source of inspiration. Unlike the Anasazi, we have built too many of our labs facing the wrong direction, away from the sun. "I hope you get your grant," I say out loud, and then I lean back against the circular wall of a ceremonial kiva, and fall asleep in the soft rays.

PHOTOZYMES

Months later, when I mention ASU's pentad light-harvesting efforts to James Guillet of the University of Toronto, he nods his head. "They're impressive, and they work well." A polite silence. "As long as you have something to plug your laser into. But what happens when you walk outside and hold them up to ordinary, northern Canadian sunlight? Can you get an electrical current? Even better, can you get fuel out of them? That's what I want to do."

And to do that, Guillet is tackling a different part of the photosynthetic machinery. While Gust and Moore are modeling the reaction center, Guillet attempts to build what he thinks every reaction center will ultimately need: a way to get the diffuse drizzle of sunlight to hit home. In plants it's done with pigment antennas, and if Guillet succeeds, the ASU team may be able to marry a full-fledged antenna to its pentad and do chemistry in water. But that's getting ahead of the story.

For now, Guillet has good reason to be seeking his own part of the artificial photosynthesis Grail. He lives in a cold, dark country that uses more energy per capita than any other in the world. Because sunny days are not terribly plentiful, Guillet is interested in finding a way to wring a storable fuel out of the sun, something to burn during those winter months—something like hydrogen. Though he can be suspenseful about his plans, I think he may be onto something. His track record—papers, awards, patents, businesses—speaks of a man who lets no moss grow on a good idea.

Though he retired from teaching several years ago, Guillet maintains an office at the University of Toronto and still comes in regularly. Here he keeps one foot in academia and another in private industry, where his career began. "I was trained in the private sector, where practical applications were king," he tells me. "But when I transferred to the university, they wanted me to give all that up." It was in an era when the real prestige in science lay in the field of physics, where elegant theories and unifying concepts were badges you could shine. Deep in the throes of "physics envy," as it's called, the head of his department actually told him he shouldn't be producing anything patentable. To Guillet's credit, he flatly ignored the advice and has been patenting inventions and fledging companies ever since.

One of his inventions is Ecolyte, the plastic that degrades into small pieces when the sun shines on it.

"I have four times as many inventions as Benjamin Franklin," he tells me, "and I'm pushing for one hundred." Somewhere between here and one hundred, he may just invent a device that spins energy from the sun into a fuel that can run your car. When it debuts, it may just sneak up on his down-south competitors. To hear Guillet tell it, he and they are in parallel races with some very different ground rules.

In the United States, says Guillet, the military approach is often employed to get really big things done, with the Manhattan Project being the model. But that won't work this time, he predicts. "Buying high-tech lasers [like the ones that zip around Neal Woodbury's train set] has always been a staple of solar energy research. But I don't think solar energy devices are going to come via big-ticket approaches. I don't think nature works at that scale."

We are strolling toward a French restaurant in the university district, and he pauses to pluck a leaf from one of the many trees lining the narrow road. "This is the solar energy device that everyone would love to mimic," he says, handing it to me like a flower. "And this device doesn't do chemistry under the concentrated, coherent light of lasers. Lasers are very intense, whereas sunlight is more diffuse—like a drizzle instead of a hard rain."

At this point he stops and squints up to the sun. "Even though a lot of sunlight falls to Earth, it is notoriously hard to collect. The trouble is timing. Green-plant photosynthesis requires that not just one but two photons hit the reaction centers of the two photosystems in rapid succession. This 'two-photon event' has to occur within the lifetime of the excited state, or the side reactions will fizzle— there's just not enough energy in one photon to drive the process." Statistically, no bookie in the world would put money on two photons hitting the same square centimeter of a leaf at almost the same time. Nature, of course, has taken these dismal odds and turned them into sure bets.

Leaves do it, algae do it. Even photosynthesizing bacteria do it. They unfurl an antenna that photons can't resist. Devoting a lion's share of their chlorophyll, photosynthesizers spread out a receiving array of pigment molecules, about two hundred for each reaction center. Each lollipop-shaped antenna molecule turns its porphyrin ring, like the face of a sunflower, toward the incoming photons. When a photon hits anywhere in the array, it excites an electron in porphyrin to a higher orbital, and before the electron can decay back to its original orbital, the energy (not the electron itself, just the energy) migrates to an adjacent porphyrin ring poised to receive the energy. "The energy migration is like the sound waves that migrate from a struck tuning fork," says Guillet. "Eventually, a tuning fork across the room will 'catch' the energy and start resonating with the same frequency."

In a leaf, the migrating energy funnels quickly down to its destination by being passed antenna to antenna. Having a whole array of these ringlike pigments on the lookout for energy is like having your whole roof collect rain instead of just the opening of the rain barrel, says Guillet. "In fact," he says, "if you hold up an antenna of two hundred pigment molecules instead of just one, you are forty thousand times more likely to have a second infusion of photon energy hit the mark when it needs to."

To do anything close to photosynthesis—to split water for hydrogen fuel using sunlight, for instance—Guillet contends we will need that second photon infusion. "Once you wean yourself off lasers, you realize you're going to need two photons arriving onstage at almost exactly the same time. No matter how good your reaction center is, it won't have anything to work with unless you can harvest photons for it." Once he faced that fact, says Guillet, he decided to let others perfect charge separation while he figured out how to make an artificial antenna.

"I wanted to see if energy would migrate along a linear chain of light-sensitive pigments the way it does along a large array. I chose naphthalene, an organic chromophore used for making dyes and solvents, because it was related to the light-sensitive parts of chlorophyll molecules. I strung thousands of these naphthalenes together in a long repeating chain called a polymer [a string of like molecules]. It may help if you think of it as a long pearl necklace on a flexible string. When I put this in solution, it coiled up. When I flashed it with light, one of the naphthalene chromophores picked up the energy, which then began to travel, not just pearl to pearl along the chain but also hopping to other parts of the chain that were coiled nearby." Guillet refers to the random hopping of energy as "the drunken sailor's walk."

Guillet also recognized that in the leaf, nature manages to gently direct this random walk—like putting the drunken sailor on a sloping drainfield so he eventually heads toward the bottom. In the plant's case, the "bottom" is the reaction-center chlorophylls, where the action really begins. Each step along the way, each antenna, is at a slightly lower level in the energy landscape. Heading from high energy to lower energy is like going down a slippery slope; the energy can't travel the other way, so it gets trapped at the central chlorophylls.

Guillet wanted to mimic nature's trick with his single chain. "After fishing the photons out of the drizzle, I wanted to have all the energy report to a single location at the end of the chain—a basin in the energy landscape." Once it was trapped in a central spot, he could devise a way to use the energy to make and break chemical bonds, to split water, to make pharmaceuticals, to do all sorts of chemistry.

Anthracene proved to be a perfect basin—Guillet put it at the end of the chain, and after the naphthalene necklace was flashed with light, the spectral signature changed, signaling that the energy had moved. "That signal was a heartwarming sight. I knew immediately that most of the light energy had left the naphthalene and landed in the anthracene. To top it off, the process was also efficient—ninety-five out of every one hundred photons of light cause anthracene to light up. This ninety-five percent conversion rate rivals photosynthesis, which told us we can build antennas which are just as good as nature's antennas are at collecting photons."

Now that you can trap the energy, I asked, what can you do with it? Guillet brightens here, and I sense that moment when a scientist wears his heart on his sleeve—for Tom Moore it was membrane potential, and for Jim Guillet, I think it's chemistry in water. "Life has some very universal, common strategies—tricks that it uses across the board because they work so well. One of these is doing chemistry in water—whether it's in a tree, a corn plant, or a brain cell—the solvent of choice is water." We, of course, have been pursuing a different tack. When we make plastics, synthetic fibers, coatings, pharmaceuticals, agricultural chemicals, and other products out of petroleum products, we use organic solvents, which can give off toxic emissions and are hard to store and dispose of safely. Once Guillet got his energy necklace to work, he started fantasizing about making these organic solvents obsolete. "I thought, why not mimic nature and use the benign fluid of life as the medium for chemistry?"

In his wildest dreams, Guillet began to see his polymer antenna ushering in a new era in which chemical-manufacturing plants would be truly plantlike. "There was only one problem," he tells me. "Naphthalenes hate water." Like membrane proteins, naphthalenes are water-fearing and won't stay suspended for very long. His solution was to attach some water-loving molecules to the chain, giving the polymer a Jekyll and Hyde personality. The hydrophillic groups would happily mingle with water, while the naphthalenes would cluster at the center, forming a cozy hydrophobic pocket.

This Jekyll and Hyde personality is a repeating theme in chemistry, even when the chemistry concerns laundry. It's because of the soap molecule's split personality, in fact, that we can get our clothes clean. Think of the soap molecule as a "magnetic" bar with a north and south pole—one pole is water-loving and one is water-fearing. Drop the tiny bars in water, and the water-fearing ends will find each other and huddle together while the water-loving ends point out toward water. Essentially you have a spiny sphere of soap molecules, called a micelle, suspended in your washing machine. At the center of each sphere is a water-fearing "pocket" that actually attracts other water-fearing molecules floating by, including greasy-stain molecules. Once these water-fearers escape from the fibers of your jeans, it's only a matter of time before they run across one of the spiny spheres of detergent. Sensing refuge, they dive into the center of the micelle and are washed away with the dirty laundry water.

The same sort of drama occurs with Guillet's new polymers. Each long, necklacelike polymer creates its own coiled mob, called a pseudomicelle, with a hydrophobic pocket in the middle. What Guillet has created is a globular antenna with a sweet spot—the energy goes to the center, and so do any water-fearing molecules in the neighborhood. When the water-fearing molecules happen to be the precursors in a chemical reaction, says Guillet, you're ready to do chemistry in water: The precursors head for the center, where they're zapped by the energy coming from sunlight, which either forms or breaks a bond.

In many ways, this is what Neal Woodbury envisions when he talks about catalysts with a power pack. Guillet has created microreactors that float in water and act like catalysts or enzymes—"grabbing" substrates in their hydrophobic hot pocket, and using the energy from sunlight to make and break bonds. He calls it a photozyme.

The photozyme that most of Guillet's studies have focused on has a nickname that numbs the tongue: PSSS-VN. It's made up of two compounds: sodium styrenesulfonate and 2-vinylnaphthalene. His first test run of PSSS-VN was in a beaker filled with water and pyrene, which is a carcinogen. As soon as he sprinkled the water-fearing pyrene into the beaker, it dove into the central pocket of the polymer coil, where it would be at the receiving end of photoenergy. When sun rays bathed the polymer, energy zipped to the center of the coil and performed extremely rapid photochemical reactions, breaking the pyrene down into less dangerous molecules.

To demonstrate his idea to a wider audience, Guillet chose something that people care about: polychlorinated biphenyls, or

PCBs. These common industrial chemicals (found in 40 percent of all electrical equipment) are now being found everywhere, even in arctic waters. The reason PCBs are so ubiquitous is that they are resistant to breakdown in sunlight. Conventional cleanup of PCBs and other pollutants is often stymied by the fact that the pollutants are present in trace amounts, spread over large bodies of water.

Photozymes offer an ideal solution because they can scavenge out PCBs, even when they are present in concentrations of only a few parts per million, and then, with the help of light energy, they can chew off the offending chlorines from PCBs, rendering them harmless. It would work like this, Guillet explains: A PCB molecule would be attracted to the center of the micelle, and once there, a shot of light energy would cause one of its chlorine bonds to sever. The micelle would then release the crippled PCB, and another would enter. In a week or two, after half a dozen dechlorinating trips to the center, every PCB molecule would be chewed to a nonchlorinated, biodegradable state.

Instead of tearing something down, I ask him, will we also be able to make something using photozymes? "Yes! You'd be surprised how many reactions can be carried out with light instead of heat or pressure or harsh chemicals. We've shown, for instance, that you can mix photozymes with the precursors of vitamin D and make it in one step instead of the several it now takes—energy courtesy of the sun. Which means, of course, a lot of energy. We figured we could make the entire annual Canadian consumption of vitamin D in a backyard swimming pool with the existing efficiency of our process."

With the photozyme, photochemistry becomes very specific— you get the product you want without the side or ancillary reactions that produce products you don't want. The process can also be calibrated. You can adjust the molecular weight of the photozyme, engineer the pocket so that only certain hydrophobic compounds can get in, or match the energy levels of the antenna to particular substrate molecules so that the antenna "finds" and excites just the right substrate in a stew of molecules. Besides being efficient and using the boundless energy of the sun, the photozyme is a durable workhorse. Once you extract the vitamin D or whatever product you are making from the solution, the polymer can be used again.

Not that "chemistry au naturel" is a complete panacea, says Guillet. It brings problems of its own, the same problems that natural organisms face. "Anytime you do chemistry in water using natural sunlight, you have to work in layers—the best light is at the top, and becomes less saturated as you work your way down. My problem—an engineering problem, really—became, how do I make sure my reactions will always be on top and exposed to maximum sunlight? I pondered and pondered this until, one day, out at my weekend place at Stony Lake, I took a walk and wound up sitting by a quiet cove. There, right in front of my eyes, was the world's best strategy for collecting light for solar-driven chemistry. I'll show it to you."

He reaches behind him for a small plastic container, then shakes its contents into my hand. Translucent plastic disks the size of hole-punches pile up in my palm. "These are Solaron beads. They're made of a cross-linked polymer called polyethylene. Right now they're dry, but if you put them in a liquid, they would quickly absorb it the way absorbent diapers do. To do solar chemistry with them, you first let them soak up a liquid starting material for a product such as vitamin D. You throw the loaded beads onto a pond, where they spread out in a uniform layer, soak up the sun, and do serious chemistry on the precursors. To remove the beads, you either screen them all at once or push them bit by bit across the surface with a slow-moving boom, letting new ones fill in behind to take their place. To 'harvest' the vitamin D, you flush the beads, then soak them in starter material again and toss them back out onto the pond.

"In many ways it resembles farming more than industrial chemical processing," says Guillet. "In fact, we can envision using grain handling equipment—airveyers, high-speed blowers, and silos—to store and transport the beads. In the company I've formed, called Solarchem, we're already making a number of products this way. It costs us about fifty cents to cover a square meter with these tiny solar-chemistry labs, in contrast to something like photovoltaic cells which cost fifty to two hundred dollars per square meter."

All this time I am rolling the disks together in my hand. Finally, I take a good look at them. They are oval and slightly concave. "Can you guess where I got my inspiration?" he asks.

I envision the tiny disks floating on the surface of water, and I can imagine how well they would pack, one next to the other, covering the entire surface. Suddenly, in a rush, I know. My quest has come full circle, and the lessons I must learn—what is a weed, what is a nuisance, what is a brilliant model of efficiency and elegance—all float to shore at once. Guillet's inspiration is the Cheshire cat that I can't catch, part of the ineluctable genius that surrounds us.

"I know exactly what this is," I tell him.

"Incredible, isn't it?" he says, and we smile at one another as he pours a small mountain of it into my waiting hands.

Artificial duckweed. Patent number 84.

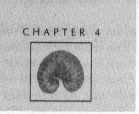

CHAPTER 4

HOW WILL WE MAKE THINGS?

FITTING FORM TO FUNCTION: WEAVING FIBERS LIKE A SPIDER

Though environmental policy makers have focused on the growing glut of garbage and pollution, most of the environmental damage is done before materials ever reach the consumer. Just four primary materials industries—paper, plastics, chemicals, and metals—account for 71 percent of the toxic emissions from manufacturing in the United States, according to the researchers. Five materials—paper, steel, aluminum, plastics, and container glass—account for 31 percent of U.S. manufacturing energy use.

—JOHN E. YOUNG and AARON SACHS, authors of *The Next Efficiency Revolution: Creating a Sustainable Materials Economy*

We are on the brink of a materials revolution that will be on a par with the Iron Age and the Industrial Revolution. We are leaping forward into a new era of materials. Within the next century, I think biomimetics will significantly alter the way in which we live.

—MEHMET SARIKAYA, materials science and engineering professor, University of Washington

"That's why babies' heads are soft," said the man riding down the escalator as I was heading up. "They haven't completely mineralized yet." Babies' heads? I ran up my escalator and joined him on the down ride. He was going where I was going.

The Materials Research Society (MRS) meeting is held every

year in downtown Boston, filling three of the major hotels to capacity. Everywhere you look there are scientists—3,500 strong—carrying their two-inch-thick book of seminar abstracts in materials science, a field most of us have never even heard of. Strange, because materials science literally touches everything we touch; every object we walk on, ride in, pick up, put on, or pour from is made of a material or several different materials. Yet the people who worry about shatter resistance, tensile strength, and surface chemistry—the ceramists and glass engineers, the metallurgists and polymer scientists—are soundly unsung. I don't know any kids who want to be materials scientists when they grow up.

Maybe the field is just too new. Materials used to be manufactured solely by nature, and we took what we were given—wood, hide, silk, wool, bone, and stone. Eventually people learned to fire slurried sand into pots and hammer iron from the Earth. Throughout history, our progress as a people has been date-stamped by the types of materials we used—the Stone Age, the Bronze Age, the Iron Age, the Plastic Age, and now, some would say, the Age of Silicon. With each epoch of civilization, we seem to have distanced ourselves further from life-derived materials and from the lessons they teach us.

In the vivid glow from the slide shows featured at Symposium S (the bio-inspired-materials segment of the meeting), I began to see that nature has at least four tricks of the trade when it comes to manufacturing materials:

1. Life-friendly manufacturing processes
2. An ordered hierarchy of structures
3. Self-assembly
4. Templating of crystals with proteins

Each of the tricks was new to me, and probably new to many of the other conference attendees who kept stopping by out of curiosity. What sets the biomimics apart from their peers is that nature's canon has become their own. If the biomimics had their way, these lessons would be the backbone of every materials engineer's education. For the purposes of this chapter, we'll take the short course.

HEAT, BEAT, AND TREAT

In the hive of the MRS meeting, forty mini-meetings called symposia are held concurrently. In each one, new findings are introduced in papers given every fifteen minutes, all week long. Most of the talks focus on the new alchemy: the synthesis of new alloys, new ceramics, new plastics made possible by impossibly high temperatures, high pressures, and strong chemical treatments. "Heat, beat, and treat" has become the de facto slogan of our industrial age; it is the way we synthesize just about everything.

Nature, on the other hand, cannot afford to follow this strategy. Life can't put its factory on the edge of town; it has to live where it works. As a result, nature's first trick of the trade is that *nature manufactures its materials under life-friendly conditions*—in water, at room temperature, without harsh chemicals or high pressures.

Despite what we would call "limits," nature manages to craft materials of a complexity and a functionality that we can only envy. The inner shell of a sea creature called an abalone is twice as tough as our high-tech ceramics. Spider silk, ounce for ounce, is five times stronger than steel. Mussel adhesive works underwater and sticks to anything, even without a primer. Rhino horn manages to repair itself, though it contains no living cells. Bone, wood, skin, tusks, antlers, and heart muscle—miracle materials all—are made to live out their useful life and then to fade back, to be reabsorbed by another kind of life through the grand cycle of death and renewal.

It was fun to watch the milling scientists from other disciplines stick their heads into the doorway of Symposium S. While most rooms at a meeting like this are resplendent with the talk of unearthly synthetics, Symposium S featured slides of coral reefs and tall trees, spruce bogs and spiderwebs brushed with dew. The latest high-tech materials here weren't experimental designs—they were ancient, biological inventions, tested and proven over millions of years on Earth. The same Earth on which we and our materials are trying to survive.

The people in Symposium S have little allegiance to the heat, beat, and treat mantra. They see the handwriting writ large—the dwindling oil reserves, the toxic nightmares of our own making, the high failure rates (breaking, cracking, stretching out of shape) of many of our materials. Despite our colossal energy expenditures, we still can't make materials as finely crafted, as durable, or as environmentally sensible as those of nature. The rhinos, mussels, and spiders in the slides all seemed to be wearing Mona Lisa smiles. Somehow, out of the world's most common chemicals, like carbon, calcium, water, and phosphate, they fashion the world's most complex materials. As any biomimic in the room could tell you, the *S* in Symposium *S* stands for *surprise*.

THE HARD STUFF FIRST

The papers presented that week split along two lines, the mostly inorganic (the hard) and the mostly organic (the soft). Nature's inorganic materials are tough, used for skeletal structure or protective armor, the shells and bones and spines and teeth of the natural world. They are crystallized versions of Earth-derived materials—chalk and phosphates, manganese and silica, even some iron thrown in for "bite." Since organisms don't produce these inorganic minerals in their own bodies, they must find a way to tempt and tame the particles of the Earth to settle and crystallize in just the right location. If you're a soft-bodied mollusk living in the rock-and-roll of the tidal zone, for instance, the best place to have a shell crystallize would be right over your head.

Oyster Envy

Rich Humbert owns a wetsuit that doesn't quite keep him warm. Even with a neoprene mask strapped over his bearded face, he must let his eyes show, and by the time he bursts up for air, his robber's mask of exposed skin is a painful shade of purple. All of which makes diving for abalones in Washington's San Juan Islands a lonely vocation.

"Most people prefer to encounter their abalones in souvenir shops," he tells me. "But I like to get in with them, see where they live." He pantomimes the hunt for me. "You reach down for them through the murky tidal wash, feeling with your hands. The outer shell is drab and scabby with barnacles. It's hard to believe that inside there's this smooth, luminous, mother-of-pearl lining. The idea is to grab them as soon as you touch them, before they can suction themselves to the rock."

A tickled abalone can be wickedly fast. So powerful is its foot suction that if you miss the magic moment, you have to pry it from the rock with a tire iron. For abalone aficionados like Humbert, a pry job is the sign of a hacker, and he would rather turn completely purple than resort to one.

Most people who hunt abalone eat the meat and sell the shell, but Humbert dives and plucks for what he can learn. He's part of the University of Washington's team investigating abalone nacre, the smooth inner coating that is delicately swirled with color and, best of all if you're a ceramist, hard as nails. "Ever try jumping on an abalone shell?" asks Humbert. "A car could drive over these guys and not faze them." Back at the lab, he has to fire up industrial machinery to break the outer shell and nacre into pieces. One shell— a beautiful eight-inch platter—will be enough to last through a year of research.

To the naked eye, the piece of nacre that Humbert hands me looks smooth and featureless. Then he shows me an electron-microscope picture of the same piece in cross section. Standing out in bold, black-and-white relief is the intricate crystal architecture that accounts for the shell's ability to shrug off stress. Looking in from the side, you see hexagonal disks of calcium carbonate (chalk) stacked in a brick-wall motif.

If you look closely between the bricks, you can see a narrow mortar of squishy polymer. The polymer acts like a thin smear of chewing gum—it stretches ligamentlike when the disks are pulled apart and it slides and oozes in response to head-on stress. If a crack does get started, the brick-wall pattern forces the crack to follow a tortuous path, stopping it in its tracks. As a result, "Abalone is twice as tough as any ceramic we know of—instead of breaking like a man-made ceramic, the shell deforms under stress and behaves like a metal," says Mehmet Sarikaya, whose name appears in the credit line for many beautiful electron-microscope pictures of abalone.

Portraits of the nacre taken from above show a further complexity. On any one level of the brick wall, the hexagonal disks are twinned: Their shapes and placement echo one another, as if a mirror is between them. Individual disks are composed of twinned "domains" that also mirror one another. Even the grains within each domain are twinned, showing the mathematical repetition and beauty that characterize natural form.

Closer to home, a soft material in our own bodies has become the poster tissue for this concept of repetition at many scales. The "unraveled tendon" drawing (which got a lot of screen time at the meeting) shows a hierarchy that is almost unbelievable in its multi-leveled precision. The tendon in your forearm is a twisted bundle of cables, like the cables used in a suspension bridge. Each individual cable is itself a twisted bundle of thinner cables. Each of these thinner cables is itself a twisted bundle of molecules, which are, of course, twisted, helical bundles of atoms. Again and again a mathematical beauty unfolds, a self-referential, fractal kaleidoscope of engineering brilliance.

In the human tendon, in the abalone shell, in the stacked plywoodlike layers of the rat's tooth—over and over again at the meeting, this issue of "structure granting function" came to the fore. The multileveled complexity of these materials is referred to as an *ordered hierarchical structure,* which seems to be nature's second trick of the trade. From the atomic level all the way to the macroscopic, precision is built in, and strength and flexibility follow.

But how does nature manage to create that microstructure? And how can we do the same? Answering those questions is at the very heart of what biomimics are trying to do. "We want to do more than just copy down the angles and the architectures of nature's designs or build our materials in their image," says ceramist Paul Calvert from the University of Arizona Materials Laboratory in Tucson. "What we really want to do is imitate the manufacturing *process,* that is, *how* organisms manage to grow, for instance, perfect crystals and form them into structures that work."

All of the materials scientists I talked to agreed with Calvert's assessment. They were itching to grow lattices with dress-parade perfection, to control crystal size, shape, orientation, and location, especially in the world of ceramics.

The ceramics we're most familiar with are glass, porcelain, concrete, mortar, bricks, and plaster, but as Paul Calvert says, "Ceramics have gone far beyond toilet fixtures and cereal bowls." They are now being used in all kinds of high-tech applications—as insulators, guides, bearings, wear- and temperature-resistant coatings, and in devices that need certain optical, electrical, and even chemical characteristics, such as sensitivity to gases or the ability to accelerate a chemical reaction. For all we are asking ceramics to do, it's ironic that we're still using Stone Age techniques to manufacture them. Basically, we take earthy inorganic particles and subject them to heat or pressure in order to squeeze them together into a substance that is hard. Says Calvert: "Our biggest problem is cracking—brittleness. In recent years, we've been making incremental progress by making our grains finer and finer. We finally have them down to the nanometer size, but we're still plagued by brittleness."

A few years ago, Calvert thought it was time to energize imaginations, so he and other biomimics began to look at natural designs. They uncovered plenty of examples of biological organisms that, like the abalone, sport hard body parts made from a mixture of inorganic minerals and organic polymers. Your bones are crystals of calcium phosphate deposited in a polymer matrix, for instance. Diatoms—those microscopic sea creatures that look like living snowflakes—have skeletons made of silica glass shaped by the organic membranes of their bodies. Teeth are inorganic crystals, as are sea-urchin spines and snail shells. The ultrahard crystals in a lamprey's "teeth" are what allow it to rasp through rock. Nature is even able to utilize some magnetic material in its mineralizing process. For instance, a bacterium that was discovered in the late seventies grows crystals of iron oxide—magnetite—in tiny vesicles (balloons) inside its body. These magnetite-filled vesicles line up like beads in a key chain, and together they help the bacteria orient down toward the magnetic center of the Earth, which is also toward the anaerobic zone where they find their food.

In all these cases, nature's crystals are finer, more densely packed, more intricately structured and better suited to their tasks than our ceramics and metals are suited to ours. The biomimics decided it was time to find out why.

Pearls of Wisdom

To understand how organisms manage this trick, it helps to understand the softer side of the composite mix. For this we have to go to the molecular level—a level smaller than Sarikaya's electron microscope can reveal. "That thin smear of polymer is more than just mortar sticking the bricks together," says Rich Humbert. "It's made of polysaccharides (sugars, essentially) and proteins, and they're the ones actually running the show." In fact, when an abalone "decides" to build nacre, the polymer mortar is erected first, and then the bricks.

This counterintuitive ordering occurs in a similar way in many biomineralizing organisms. First the organism's cells secrete proteins, polysaccharides, or lipids (depending on the species) into the fluid surrounding them. These "framework" polymers self-assemble into three-dimensional compartments (cubes, rectangles, spheres, or tubes) defining the space that is to be mineralized. "You can think of the framework polymers as the walls and ceilings and floors of a room that will eventually be infilled with mineral crystals," says

Humbert. In the case of the abalone, the organism builds not just one room but an entire apartment building, laying down one story of rooms after another, each slightly offset from the one below to accomplish the interlocking brick-wall motif.

Inside each room is seawater saturated with calcium ions and carbonate ions—charged particles that will eventually land and aggregate into a crystal of calcium carbonate (chalk). Because the ions are charged, they don't just randomly precipitate out of solution—they are attracted to oppositely charged chemical groups protruding from the walls of the rooms. Once that first layer of ions settles out, it sets the tone for the rest of the crystal. Like the bits of dust in the supercooled beaker of your high school chemistry lab, the first ions will act as seed kernels or nucleators, and the rest of the ions will settle around them, growing a crystal of a particular shape. Since the crystal's strength and function depend on shape, the ions' landing locations turn out to be key.

The mollusk, evolutionarily eager to build a shell of herculean strength, found an ingenious way to get those ions to settle into a particularly strong shape. Here's how it works: After the framework of rooms is assembled, the mollusk releases templating proteins into the inner rooms. These proteins self-assemble into a "wallpaper" that peppers the room with an orderly array of negatively charged landing sites. If we were the size of atoms, we could walk among the chemical groups and feel their electrostatic pull, beckoning to positively charged ions in the seawater, such as calcium.

To visualize the proteins in this special wallpaper, a quick biology lesson is in order. Proteins (which make up 50 percent of the dry weight of every living cell) are large 3-D molecules that begin as long necklaces of dozens or even hundreds of chemical groups called amino acids. Each amino acid has a different constellation of charges, and when the chain is released into the fluid of the cell, those charges cause the protein to fold up in a very particular way.

The folding pattern has a lot to do with how the amino acids take to water. Neutral, water-fearing amino acids will burrow into the center of the protein complex, while the charged, water-loving ones will take to the periphery. The amino acids also interact with one another—some repelling their neighbors and straining to get away, others meeting in a bond. What results is a three-dimensional shape, a form uniquely suited to its function. A protein may have a structural role in the body, assembling into tissues and skeletons, or it may have a "trade." Hemoglobin, insulin, neuron receptors, antibodies, and en-zymes (which orchestrate and speed up chemical reactions) are all proteins, plying a particular trade based on their shape.

In the case of the templating proteins of abalone, the protein chain folds into a zigzag shape which bonds side by side with other zigzagging proteins to form an accordion-pleated sheet (the wallpa-per). There are two "faces" to this sheet—some groups of amino acids stick out into the room, while others are embedded in the walls, floor, and ceiling like anchors. Daniel Morse, director of the Marine Biotechnology Center at the University of California in Santa Barbara, has determined that the groups that anchor in the walls are neutral (principally glycine and alanine) and those that stick out into the room are negatively charged (principally aspartate).

The landing sites on the pleats are not random, either. Because each zigzag protein is itself precisely formed (templated by DNA), its amino acids are studded along its surface predictably. Every few nanometers, they sit ready to snag oppositely charged ions floating by in solution.

How the ions are arrayed in that first layer says everything about how the crystal will look and function. One pattern may yield rhom-bohedral crystals like those in nacre; another will yield prismatic crystals like those in the abalone's hard outer shell. Different shapes and orientations and sizes determine whether the crystal will have optical qualities, be able to conduct electricity, or be hard or soft. There are fourteen different shapes of crystal that are possible in all of nature.

Now, what if we were able to template for any one of those fourteen kinds of crystals by using different protein craftsmen? What if we could coat an object with a film of proteins and then dip it in seawater and have nacre self-assemble into a hard coat? That's the dream, and there's one fact that makes it possible—proteins don't need to be in a living cell to do their thing.

A protein separated from a living cell is still a protein—fully charged and able to direct crystallization. In fact, that's what happens in the abalone—the proteins are pumped outside the cells into a seawater-filled gap between the soft body and the harder outer shell. That means, theoretically, that we should be able to fill a beaker with proteins and seawater and watch as the proteins self-assemble into their rooms and their wallpaper and the ions nucleate and begin to grow into crystals.

Self-assembly, then, is nature's third trick of the materials trade. Whereas we spend a lot of energy building things from the top down—taking bulk materials and carving them into shape—nature does the opposite. It grows its materials from the ground up, not by building but by self-assembling.

Self-assembly rides the riot of forces ruled by classical and quantum physics. Like charges repel like charges, but opposites attract. Weak electrostatic bonds hold molecules together gingerly, and as conditions change, they can easily correct and adapt. Stronger, more permanent bonds are consummated with the help of lock-and-key catalysts called enzymes.

Before any kind of bond can be formed, however, wandering molecules must first collide, like guests at a cocktail party. The energy that keeps molecules mingling comes from what scientists call Brownian motion, named after Robert Brown, an early-nineteenth-century botanist who asked the world, "Have you ever noticed that pollen grains stay suspended in water all by themselves?" (In those days, an observation like that could make you famous.) A generation later, Albert Einstein explained that the pollen grains are buoyed by the fact that invisible water molecules are continually knocking into and moving them. This restless bumper-car action of molecules also occurs in air, which is why dust particles look as if they're dancing in sunbeams.

Once molecules collide, those that are shaped peg to hole like Lego blocks snap dutifully together. All of this assembly, unlike our building of materials, is energetically "downhill." It's order for free. Proteins are amenable to this sort of self-assembly because of their shapes and their "electric" personalities (how their charges are distributed). These precise qualities are set forth by genes—informational templates that contain the code for making proteins. Once gene-templated proteins self-assemble into their accordion sheets, they themselves become the templates for making exquisite shells. The templated becomes the template.

Which leads to nature's fourth trick of the trade—*the ability to customize materials through the use of templates*. Whereas we muddle by in our industrial chemistry with final products that are a mishmash of polymer-chain sizes, with most too long or too short to be of ideal use, nature makes only what she wants where she wants and when she wants. No waste on the cutting-room floor.

If we want to emulate nature's manufacturing, we have to get backstage and interview the proteins, those templaters that make precision assembly possible at body temperatures. We have to learn their amino acid sequences and figure out how to produce them in commercial quantities. With the help of these "invisible hands," the biomimics hope we may be able to sculpt with geometric precision, and do away with "heat, beat, and treat."

The Great Protein Sequence Hunt

Mehmet Sarikaya's eyes, the color of Turkish coffee, flash a warning to each member of the biomimicry team. "Before we do anything, we've *got* to find the protein sequence." He is literally straining with impatience, determined to be part of the first team to find that protein-sequencing data. "We are not the only lab working on this," he confides to me at a harried luncheon meeting, "but we are the only ones on the right track." As he describes it, the race for a test tube full of honest-to-god, framework-and-wallpaper proteins is furious, and Sarikaya, elbows flailing, wants to win. I briefly imagine him crossing the finish line and renaming the field Biomehmetics. Later, when I tell my joke to someone who works for him, they say they are sure he has already proposed it.

Right now, Sarikaya is on the warpath because he feels the team is stalled. I am attending a preparatory meeting for an upcoming science conference at which team members will present their work. Rich Humbert, the abalone diver-scientist, is showing pictures of his latest experiments. So far, Humbert has managed to get a random mix of abalone proteins to form "artificial pearls" against the side of a test tube. When the pearls are cut open and magnified, you can see protein (stained orange) crowded into circular layers. This layered "jawbreaker" doesn't have the exquisite brick-and-mortar architecture of real nacre, but at least it implicates protein in a supervisory role. This has plunged Humbert deep into speculation about how nacre development might have evolved, and he would like to write a paper about it. Sarikaya fumes about the time it will take.

He wants Humbert to find the abalone proteins responsible for nucleation, so the team can attach them to the surface of an object, dip the decorated object in seawater, then watch the nacre crystallize. The sooner the better. The military is equally interested in this idea of stronger coatings, because it, like the abalone, is often in zones of serious insult and injury, where fracture resistance would be a virtue. To that end, the Office of Naval Research has awarded a three-year grant to the UW team to investigate abalone shell, a study of what they call "layered nanostructures."

The team at the University of Washington is wonderfully inter-disciplinary, and it is here that I see the future of biomimicry. Engineers and materials scientists are working alongside microbiologists, protein chemists, geneticists, and Renaissance thinkers like Clement Furlong.

If there is a counterpoint to Sarikaya's intensity, it is Clem Furlong's ease and patience. Furlong is Rich Humbert's supervisor and leader of his own department in medical genetics. Deep in the maze of a huge building, I find him shoehorned into an office that threatens to collapse around and on top of him. Papers are stacked atop filing cabinets all the way up to the high ceilings. Tables are heaped with journals from half a dozen disciplines, and computers lie about in various stages of undress, their circuitry hanging like mattress stuffing. Furlong and his students have just built five computers from mail-order parts this week, and he is positively gleeful about how easily one can assemble a Ferrari of a machine. He finds a piece of blank paper (no small task in that office) and writes up a parts list for me, with exact prices from memory, as if he were writing a recipe for his favorite hors d'oeuvre. For Furlong, I suspect, science is a way to get paid for tinkering.

Somewhere in those stacks—he points to the dusty neighbor-hood near the ceiling panels—there are patent certificates for Furlong's inventions. He has a hefty vita as well—lots of papers on medical genetics—but he seems most proud of the things he has made. A new Furlong invention, in fact, may be instrumental in the team's quest to mimic abalone shell.

"Once we sequence the protein," he says, "we'll have to find a way to produce lots of it. We can't continue chopping up the shells." Besides the risk of overcollecting the species, the grinding is hard on the proteins—it either truncates or destroys them.

An alternative would be to conscript the trusty *E. coli* bacteria (found in the human gut) to make those proteins for us. It wouldn't be the first time we harnessed bacteria to help us make products. For thousands of years, we have used yeast, bacteria, and molds for brewing beer, making wine, leavening bread, and culturing cheese. Today, bacteria grown in vats are persuaded to produce food addi-tives, antibiotics, industrial chemicals, vitamins, and more. We have

bred the tiny microbes like livestock, customizing them through artificial selection.

There's a difference, however, between this kind of bioprocess-
ing and the modern version, called biotechnology. With biotechnology, we genetically alter a bacterium's manufacturing processes by
splicing in a gene from another species. To make insulin, for instance,
we take the human gene for insulin manufacture and splice it into
E. coli. By cutting and splicing, genetic engineers assure me, they are
simply imitating a technology that bacteria themselves have long
practiced. Genes from one species of bacteria are freely transferred
to completely different species of bacteria. That's how the global
microcosm has been able to adapt so quickly to cataclysmic change.
But human genes to bacteria? Abalone genes to bacteria?

No matter how many times I hear scientific assurances of safety,
I can't shake the feeling that it is the height of hubris for us to cross
that interphylum line, to take a gene from one class of animal and
insert it into another. I tell them I would be more comfortable if we
could culture whole cells from the abalone in a vat, and milk protein
from those cells. For many reasons, they tell me, this is not yet prac-
ticable.

So I am left with a dilemma that cropped up often in researching
this book. Counterbalancing my real fear about genetic engineering
is my real desire for us to find more benign ways of manufacturing.
With my ears open and my caution up, I learned what I could about
this technique, all the while hoping the problems with cell culturing
would be ironed out soon.

Once the protein is sequenced (soon, says Humbert), the dip-and-
coat procedure for making nacre will be halfway home. Knowing the
protein's makeup, team members will use a machine to synthesize a
segment of DNA that is the recipe template for "how to make nacre
protein." They'll insert this DNA into *E. coli,* and hope for the best.
With luck, the *E. coli* will follow the coded instructions and use its
own cellular machinery to manufacture the proteins to order. It will
essentially be a farming operation, where bacteria, like so many milch
cows, produce a continuous stream of ceramic-crafting proteins.

That's where Clem Furlong's latest device will come in handy.
Furlong's bioreactor will house the *E. coli* and provide them with
food, water, and air, thus automating the production of proteins. The
prototype bioreactor looks like a small shoebox with glass walls. Ten

or twelve transparent partitions slide into the box like slices of bread in a loose loaf. On each glass partition there are thousands, millions, of immobilized *E. coli* capable of producing one perfect protein after another. A flow of liquid nutrients surrounds them, and oxygen bubbles up from the bottom.

As Furlong explains, "The same flow that carries in nutrients will, at the other end of the box, flush and carry off the protein they are producing. This protein—call it protein A—will flow into a beaker. But say you wanted an assembly of two proteins. You could engineer one strain of *E. coli* to produce protein A, another to produce protein B, and then place them in fifty-fifty proportions on the glass slides. You'd then have proteins A and B flowing into solution, finding one another, and self-assembling in the beaker. Want a different combination of proteins? Put a different slice of protein factories in."

The proteins can be anything the biomimic might imagine— proteins that would nucleate an even harder coating than abalone, or perhaps a thin film of crystals with electrical or optical qualities. While Furlong dreams of how we might use the bioreactor, Humbert and company are trying to find the abalone proteins that will take the shakedown cruise.

Rich Humbert describes this protein identification, sequencing, and cloning strategy as if he's telling me how to cook a roast. First you extract a stew of proteins from the intervening layers of the nacre and try to separate out and identify as many proteins as you can. Most of them turn out to be insoluble (they won't stay dissolved in solution), and as such, they aggregate at the bottom of a vial and can't be separately named. Those that do dissolve in an acetic acid solvent are all you have to work with; to separate them, you first run them through an electrified gel.

To prepare for this gel electrophoresis, you add detergent to the proteins, which neutralizes their charges and equalizes their shapes. You then pour the soapy proteins near the top of a slab of polymer gel and throw the switch, shooting an electric charge through the gel. This starts the proteins shimmying down through the gel, moving at different speeds depending on how heavy they are (the lighter they are, the faster they are). After a while you see a banding effect as the proteins settle to certain locations in the gel.

Each band represents a different protein. You transfer these bands to a paperlike sheet and literally cut out the bands of purified proteins or fragments of proteins and place them in separate vials. Then you take each vial and expose the proteins to another lab technique called protein sequencing. Using enzymes that are specially designed to chew off one amino acid at a time, you figure out the lineup of amino acids in each protein. Then you congratulate yourself, take a deep breath, and put on more coffee, because you've still got a ways to go.

Fishing for Templates

One of the key discoveries in molecular biology is the procedure that enables scientists to find the gene or the portion of a gene that is .responsible for producing a particular protein. Like contestants in a game of *Jeopardy!*, gene hunters work backward. They are given the answer—protein—and they have to find the question that would have generated that answer.

That question—the code for protein—is a carefully crafted DNA segment sitting in the cells of abalone. To find this particular strand of nucleic acid in the huge abalone genome, you make yourself a probe: a piece of DNA that will match, and stick to, the DNA you want to find.

You can make a DNA probe from scratch using a machine that automatically strings together designated sequences of nucleotide bases, the subunits of DNA. You simply dial up an A (adenine), T (thymine), G (guanine), or C (cytosine), and the machine drips the base out of a vial and welds it to the end of a growing string called an oligio. (What's amazing to me is how scientists know *which* bases to dial to code for a particular protein. We know, Humbert explains, because we know the DNA code representing each of the twenty common, natural amino acids that occur in all life-forms. This genetic code, one of the truly amazing findings of our time, is simple enough to be printed on a 3-inch by 3-inch chart. Most labs keep it taped right on the oligio machine.) Using other beguilingly simple genetic engineering techniques, you make millions of copies of this probe. Now you're ready to go fishing.

The other part of the process is building a fishing pool of segments of complementary DNA (cDNA) that you derive from the abalone. This process is called making a cDNA library. From one of the big scientific supply houses, you order a kit that essentially takes the tissue from the abalone and transforms the messenger RNA found in the cells into complementary DNA. Then you go fishing in this pool of cDNA, trolling until your DNA probe finds a complementary strand and sticks to it.

The matchup is possible because of the laws of complementarity. That is, if you have a base A on your DNA strand, it will always match up with a base T on the cDNA, a C will always bond with a G, and so on. Chances are, your relatively small fishing probe will hook onto a much larger segment of cDNA, thereby calling attention to the whole gene—the one that holds the instructions for how to make an abalone shell protein. If all this works, you fish out that abalone gene, convince an E. coli to accept it, and cross your fingers in the hope that it will produce, or "express," the protein for you.

To find out whether the E. coli has cooperated, you need some way of seeing which colonies (out of thousands spread onto petri dishes) are producing abalone protein. The best way to do this is to go fishing again with another biological probe, this time a molecule that excels at recognizing proteins: an antibody. Our immune system produces antibodies by the millions when we are invaded by a foreign molecule. Like attack troops, the antibodies recognize this foreign object by its shape, then glom on and interfere with its functioning. What Humbert and company need are antibodies that will glom on to the shell proteins in a plate of E. coli. For this trick, they pull out a rabbit.

After Humbert purifies the protein from nacre, he will inject some of this mollusk protein into a rabbit. The rabbit's immune system, unused to mollusk proteins, will see them as foreign and create antibodies shaped to fit them. Humbert will then extract these antibodies from the rabbit's blood and modify them so that the next time they attach to a protein, the attachment will trigger an effect that Humbert will be able to detect with his instruments. Thus labeled, the antibodies are then spread onto the dishes of E. coli, and if abalone proteins are anywhere on the plate, the antibodies will head right over and stick to them. Using instruments to detect a "score," Humbert can then pluck out those E. coli colonies that are expressing the mollusk protein and let them reproduce to their heart's content. They and their offspring will be the new tenants of Clem Furlong's condo by the sea—the bioreactor.

But what happens when we do find a way to produce abalone proteins to our heart's content? Will our crystals grow as well as abalone's do? Can we use slightly different proteins and produce slightly different, custom-made crystals? These questions can be answered only by going through the motions—setting out proteins or protein analogues and letting them grow crystals.

Growing Crystals Nature's Way

Galen Stuckey, Department of Chemistry, and Daniel Morse, Department of Molecular, Cellular and Developmental Biology, University of California at Santa Barbara, have learned as much as they need to know about abalone proteins, and they're moving on. Like the Washington team, they found it difficult to break the stalemate of insoluble proteins, those that lump together in the bottom of the beaker instead of yielding to water. Even those that could be dissolved rarely revealed their complete amino acid sequence. Rather than wait for a complete sequence, Stuckey and Morse decided to bank on the one large clue that kept turning up: the preponderance of *acidic* amino acid groups in all the proteins they could measure. They made themselves a protein analogue—a simple chain of acidic amino acids—as a stand-in for the real thing.

Hoping to see mineralization in the act, they first had to convince the protein analogue to embed itself on a surface that would act like the walls and floors and ceilings of the abalone's scaffolding. The surface they chose is called a Langmuir-Blodgett, or L-B, film. Basically, it's a slick of tadpole-shaped molecules that float atop a pan of water. Each molecule's bulbous head is a charged group and the fatty tail is neutral. Because water is slightly charged, the charged head is attracted, while the neutral tail is repelled. To create an L-B film, these molecules are spread onto a shallow tray of water, and then herded together by a boom that moves across the surface. The boom actually squishes the molecules together until they "stand up"—with water-loving heads buried in the surface, and tails extending above. In the cartoon sketches that scientists have drawn for me, an L-B film looks like a putting green of grass blades.

To get crystals to grow from this ceiling of molecules, Morse pours some zigzag, accordion-sheet proteins into the tray of water. With the help of chemical hooks, the neutral side of the protein sheet embeds itself in the fatty film ceiling, while the negatively charged pleats hang down into the water, creating a wallpaper of landing sites, just as in the abalone's "rooms." He then adds mineral ions to the water and lets crystals grow like stalactites from the ceiling. By being able to control the placement of the nucleation sites, Morse has found he can essentially direct what kind of crystal will form. He is now on to stage two, trying to identify the "pruning" proteins that are also present in abalone, thought to float around in the abalone "rooms" and terminate crystal growth.

So far, Stuckey and Morse have used only calcium carbonate (chalk), the choice of abalones. Other biomineralizers in nature (the sixty species that have been found so far) are known to work with many more exotic materials. Curious about these other materials, Peter Rieke of the Pacific Northwest Labs is going out on a ledge.

Crystal Windshields

Peter C. Rieke, mountain climber and materials scientist, takes both his recreation and his science to the edge. When I visited him at his Richland, Washington, lab, he was bundled in a three-blanket head cold that he caught while hanging against a rock face one snowy night in Yosemite National Park. The next time I saw him, half a year later at the Boston MRS meeting, he and his wheelchair were being hoisted onto a speaker's platform that was not handicap-accessible. He had broken his neck and other bones in a climbing fall that should have killed him. When he greeted the MRS conference crowd with the customary "I'm glad to be here," he paused a beat and then added, "believe me."

Like Morse, Peter Rieke is also trying to grow crystals on a thin film, but instead of using L-B films, he's trying lab-made films called SAMs, or self-assembled monolayers. Instead of being perched on the water's surface, SAMs are films that coat glass slides at the bottom of a tray of solution. Instead of adding wallpaper to the film the way Morse and Stuckey's method does, the charged chemical groups in SAMs are part of the film itself. That gives Rieke the ability to play with SAMs the way a mosaic artist plays with tile. "When we create the film, we can place our functional groups wherever we want them, presenting a mosaic of positive or negative charges to the ions," he says. The ions touch down on these landing sites and crystals bloom from them. "Ultimately, we'll be able to grow several different types of crystals on the same patterned film."

Though Rieke's work takes its inspiration from the organic templating of seashells like the abalone, he admits that it's not nearly as complex. "It's important to remember that with thin films, we're still working in only two dimensions," he says. "Whereas nature builds a whole apartment complex between the abalone body and the outer shell, we're just building a crystal sheet—like the braided rugs in those apartments."

In Rieke's lab I see some of the first experiments, which, despite the groundbreaking work that went into them, are deceptively humble-looking. They are simply glass microscope slides that have been dipped in a coating of polystyrene substrate, the same stuff used to make squeeze bottles, bottle caps, and drinking glasses. Rieke uses polystyrene as his substrate because it's a polymer (a repeating chain of styrene molecules), analogous to the biopolymer sheets that mollusks use. He's "decorated" the polystyrene with sulfonate groups, similar to the acidic sulfate groups associated with nucleation in mollusks. In his spare time, Rieke has experimented with other substrates and a half dozen functional groups associated with other hard-bodied creatures. The mineral ions he's paraded past these groups include lead iodide, calcium iodate, and iron oxide, in addition to good old calcium and carbonate.

In the real world, these humble-looking thin-film coatings could have a variety of applications. General Motors funds part of Rieke's research because it is interested in hard, transparent coatings for the windshields of its electric cars. "One of the reasons we aren't driving electric cars," says Rieke, "is because we can't find a way to seal in heat and air-conditioning, which escape through the lightweight plastic windows. Right now it takes too much energy to keep the cars comfortable *and* power their engines. If we could find a way to insulate the windows with a thin film, it would remove a big stumbling block from that technology."

Car companies also need coatings for their drive gears, preferably an abrasive substance that is as thin as a second skin but will not wear down. Coatings now applied to these many-faceted gears are essentially spray-painted on in a technique called "mass transfer limited." It is literally limited in that the spray doesn't reach all the nooks and crannies of the gears. "What would be ideal," says Rieke, "is if we could dunk the plastic parts in a solution of organic molecules which would adhere to every nook and cranny, and then dunk the part in a concentrated solution of precursors for an abrasive mineral. The organic molecules would act as attractors—nucleating sites for crystallization—and you'd wind up with a highly dense, perfectly oriented and ordered thin film." The same sort of film could be used to line featherweight plastic fuel tanks and parts for electric cars.

Besides abrasion- or corrosion-resistant protective coatings, whisper-thin films are also coveted by industry for electronic, magnetic, and optical devices in which precise and tiny crystals are needed to store, transport, or relay signals of light or electrons. Because they are so thin, the films could be built up into multilayered devices composed of a semiconductor layer, an oxide dielectric layer, a magnetic layer, or a ferroelectric layer for electro-optical devices. Depending on what kind of mineral you use, you could also use the crystallized coating as a sensor, a catalyst, or even an ion-exchange device.

A simple, two-bath dunking—first in template molecules, then in a bath of crystal precursors—would be a liberation from today's slow and expensive methods of producing high-density precision films. "Nature's idea of mineralization templated by proteins would revolutionize thin-film technology," says Rieke. Even something as simple as an audiocassette or a computer disk could be vastly improved. Iron-oxide crystals, common in magnetic bacteria and in gastropod teeth, are what hold the zeros and ones in our magnetic media. Right now, they are essentially piled onto the surface in disarray. Lassoing and roping these crystals into alignment with protein templates would allow more crystals to fit on a disk, holding more bits and bytes.

Ultimately, Rieke's team hopes to build a catalog of mineralizing systems, showing which crystal grows on which substrate in which concentration. "We're learning the principles of crystallization as we go along," he says, "but it's still very much a black art. It took us three years of fiddling to learn the iron-oxide system, but now that we have the recipe down, no one else will have to reinvent it. In the future, materials engineers won't have to start from scratch every time they need a two-dimensional coating. They'll just buy a kit and read the instructions: 'Use this SAM in this concentration of this solution for this long.' "

Three-Dimensional Crystal Containers

But why stop at two dimensions? Stephen Mann, a biomineralization expert in Bath, England, is re-creating three-dimensional protein sheathing, using tiny balloonlike compartments to mineralize small particles. His inspiration comes from the vesicles that living cells use to trap ions and precipitate out minerals. One-celled magnetotactic bacteria, for instance, produce incredibly tiny, defect-free crystals wrapped in organic membranes. Engineers can think of any number of uses for such small, perfectly formed, independent crystals. For instance, when you use magnetite as a catalyst to speed up chemical reactions, you would rather have a million small, separate spheres (with a lot of surface area exposed to the reaction) than a hundred large spheres. Unfortunately, without being pre-organized in balloonlike separators, most processed magnetite winds up sticking together because of the magnetic force between the particles.

To remedy this, Mann has followed the bacteria's lead, successfully growing crystals in lab-made vesicles. He's even built his organic balloons in various sizes and shapes, showing that curved, organic surfaces can also help us shape tiny single crystals with precision. Recently, Mann has utilized an even smaller compartment formed by a single cagelike protein called ferritin. (Ferritin is the protein that sequesters iron oxide in our bodies, thus keeping rust out of our cells.) Growing a crystal inside one protein would take templating to a new high (which, sizewise, is a new low).

Another way to "grow" a three-dimensional crystallized structure is to begin with a quivering block of jellylike polymer studded with inorganic minerals. As the jelly sets, the minerals inside crystallize, and the result is a composite—a flexible polymer stiffened by swarms of inorganic crystals. The combination of hardness and flexibility, say the materials scientists, would come in handy in everything from aerospace to appliance design. Imagine a living-room window that is as rigid as glass, yet able to bend and bounce back when assaulted by your neighbor kid's baseball.

Right now, we can create composites only by placing the fibers or crystals layer by layer, which is slow and expensive; crystals growing on their own inside polymer would allow us to create readily moldable composites (like car bodies) with a dramatic reduction in production costs and pollution.

Fabbers

What if you want a three-dimensional material that has an even more exacting crystalline order? What if you want a whole computer monitor, say, made of crystals in brick-wall architecture? That's a job for 3-D templating, say the scientists, using proteins that will self-assemble into a scaffolding. In the meantime, for those who still want to put nature's blueprints to work, there's a halfway technology that could give us a taste of future complexity. It's called free-form man-ufacturing, and with the aid of computers, it allows us to build 3-D objects from the ground up, one layer at a time.

Engineers have been using this technology for years to build plastic prototypes from design sketches. They take a design, digitize it in three dimensions with CAD (computer-assisted design) soft-ware, and then electronically slice the design into very fine cross-sectional layers, like those you see in magnetic resonance imaging (MRI) scans. Each slice is a complete blueprint for that layer—in-cluding its dimensions and what material it should be made of. The software sends these coordinates to the ink-jetlike heads of a rapid prototyper, or ''fabber,'' which will ''print'' the object from the ground up, layer by layer, until a three-dimensional finished product is built. Instead of ink on paper, the heads shine a laser beam onto the surface of a vat of a liquid polymer that hardens in the presence of a laser. Here's a description from the ''fabber page'' on the Internet.

> To print, say, a coffee cup, a fabber trains its computer-guided laser beam onto a vat of the liquid polymer. The laser first scans a solid circular region on the surface of the liquid, hardening it into a disk—the base of the cup. Then that base, which rests on a platform in the vat, is lowered about five thousandths of an inch, just enough for a thin film of liquid polymer to wash over it. The laser traces a hollow circle over this liquid, forming the bottom layer of the cup wall, which fuses with the base. Layer after layer, the laser traces the cross section of the cup, building it from the bottom up—including the handle. By printing one cross section at a time, a fabber can build objects that are much more complex than a coffee cup.

For the biomimics who study shell- and teeth-building technologies, the fabber's moving-front technique is familiar. Nature's twist is that instead of just one material, two or more may be used—a layer of chalk separated by a layer of proteins, for instance. Paul Calvert is now working with a company in Arizona to retrofit a fabber so that he will be able to build bio-inspired composites of more than one material.

Paul Calvert loses his normal nonchalance when he talks about the possibilities. ''A layer of templating proteins may be laid down, for instance, and then along that front, a layer of mineral precursors could be laid down. We could use ink-jet heads to deliver the material. Crystals could be allowed to grow naturally, or they may be treated in some way to accelerate growth. The next layer could be composed of an entirely different mineral." Even within a layer, a mixture of two or more materials could be used, allowing you to blend from one material to another in a gradient. "A gradient of one material to another makes for a stronger joint and eliminates the need for glues or snaps. Nature uses blurred boundaries all the time, avoiding abrupt interfaces, which are crack prone and require some kind of fastening together," says Calvert.

This kind of layerwise growth should also give engineers the ability to vary the dimensions within a part, just as bone varies in orientation and density throughout its length, becoming thicker and thinner in places. Using the fabber, we could conceivably follow nature's design plans much more closely than we have ever been able to do.

For now, Calvert and his company have not attempted anything more complex than some rings and cylinders made of two materials, and once, a high-tech Easter bunny figurine for an April display. Easter bunnies built layer by layer in 3-D might not constitute a materials revolution, but airplane wings or car bodies just might. Imagine being able to make light, strong composite skins for solar-powered cars without the use of high heat or chemicals. Or being able to fashion a spare part for your car when you are in a remote area, using common materials like chalk or sand. Sound like *Star Trek?* Stay tuned. With nature's blueprints and Paul Calvert's machine, science fiction might just materialize into fact.

THE SOFTER SIDE OF MATERIALS SCIENCE— HIGH-TECH ORGANICS

Of all the materials made by biology, minerals star in only a portion. Life has also created a bounty of resilient, organic materials—skin, blood vessels, tendons, silk, adhesives, and cellulose, just to name a few. At the MRS meeting, the fans of these organic tissues gave the biomineralists a run for their money.

Not that the two groups were far apart when it came to nature's trade secrets. Like biomineralized structures, organic materials are also hierarchically ordered. Their structure is just as faithfully coupled to function. They are templated to order, and they are self-assembled at life-loving temperatures and pressures, with no toxic aftertaste.

The only difference between the soft and the hard is where the precursors or building blocks originate. When a bombproof covering is required, inorganic minerals from the Earth come to the rescue. But when something more flexible is needed, life can build every bit of it from organic (carbon-based) building blocks. Here, proteins become more than directors or scaffolds; they actually *are* the material.

To find this softer side of materials science, I traveled to the salty tureen of life on the other coast to see how a small blue mussel uses a waterproof adhesive to tether itself to solid objects in turbulent tides. University of Delaware researcher J. Herbert Waite, tenacious in his own right, is happily stuck on *Mytilus edulis*. After thirty years of study, he's begun to pry loose the secret behind the real, live superglue made from protein.

Byssus as Usual

"We have Batman and Spiderman," yells Herb Waite at the top of his voice. He is yelling because the Atlantic breezes in December are fierce and we are out on a pier in the marsh grasses, kneeling beside a rusting fishing boat owned by the University of Delaware's Marine Sciences lab. "But mussels are every bit as talented. I can't believe we have no mussel superheroes."

Waite wears a British driving cap, and a full beard and broad chest à la Hemingway. He is reeling up something heavy, pulling hand over hand on a thick and slimy rope. Finally the dark waters part and a four-foot-wide cage comes up, its sides encrusted with navy-blue bivalves called *Mytilus edulis*, common both to salt marshes and appetizer menus. (I am glad now that we declined to order them at the restaurant where we had lunch. We were talking too highly of them to start dipping them in drawn butter.)

"How do you suppose they are hanging on?" he yells, and I realize I don't know bivalves well. I look closely and begin to see hundreds of small translucent threads, extending like plastic tethers from the bivalves to the cage.

"Those tethers are called byssus [pronounced *biss-us*], and they're more amazing than anything you can imagine. There's four or five patents right there that industry would love to have." Thankfully, Waite agrees that it's too cold to be standing here staring at gaping bivalves. We drop the cage and run back to the Cannon Hall marine lab, a building that looks for all the world like a ship gone aground. It even has porthole-shaped windows.

Once on board, we head for the tanks, where Waite has hundreds of *M. edulis* growing. Through the glass, we get a close-up of the translucent threadlike filaments—about two centimeters long—extending from the soft body. At the end of each filament is a tiny disk, called a plaque, attached to the glass with a dab of natural adhesive.

Waite sticks his hand in the tank and dislodges a few mussels from their tethers so we can watch them create new ones. "When a bivalve wants to settle down somewhere in the tidal zone to feed, it sticks out its fleshy foot [which looks more like a tongue] and creates one of these thread-and-plaque-and-adhesive combos," he says. The whole thing is called the byssus complex, and its manufacture is nothing short of fantastic.

The fleshy foot presses tip first against the attachment site. Specialized glands secrete collagen protein (the same protein that's in our tendons) into a longitudinal groove in the foot that acts as a cast or mold. The thread and plaque self-assemble and harden in the groove, and then an adhesive gland near the tip of the foot squirts adhesive protein between the plaque and the surface. The entire process, including curing of the adhesive, takes only three or four minutes.

Depending on the shear of the waves, a bivalve may put out two or three more tethers, all directly opposing the stresses. Once it's staked down, it can gape open its shell and do the filter feeding that makes turbulence a friend. Tidal flows are like a conveyor belt, sweeping in food and sweeping out wastes. Even gametes—reproductive cells—are delivered and swept away by the tides, enabling mussels to date and mate over long distances. With byssus, says Waite, mussels build themselves an anchor, a lifeline, and a niche.

It's no different from what we do. "Nature invents and we invent. In fact, I think that humans and all other life-forms have been evolving toward similar points, but other organisms are simply farther along than we are. They have already faced and solved the problems we are grappling with. For instance, *edulis*, wanting to eat in the tidal zone, had to manufacture a glue that could stick to anything underwater. We know how tough that is, because our adhesive industry has been struggling for years to come up with an adhesive that can work in moist conditions and stick to anything. It's still out of reach. Mussels are light-years ahead of us."

To prove his point, Waite gives me a primer on primers. We prime before we paint because we hope it will help the paint stick a little better. But our primers are notoriously unreliable. Water eventually eases its way under both the paint and the primer, bubbling our house paint and spreading a rash of rust across our trusty Toyotas. Water is also the enemy in the application stage, which is why we always have to dry off a surface before we glue anything to it. That's also why we have to dry-dock our boats to repair them, and why we have to use stitches in surgery instead of glue. We are flummoxed by the fact that crafty mussels are able to spread adhesive in the deep, cure it wet, and then count on it to stick to just about anything, all while *surrounded* by water. How do they do it?

"They do it with chemistry," says Waite, "and I became obsessed with finding out what kind of chemistry." I look through the glass, but the byssus-building mussel "plays poker," hiding most of what it is doing inside its fleshy foot. Waite has used molecular probes and other ingenious techniques to spy on each part of the process. As my interpreter, Waite explains what he thinks is happening inside the foot, and what we would do if we were attempting a similar feat. It's the classic "them and us" story that biomimics are so good at telling.

Cleaning the Surface

"OK," Waite says, "pretend I'm *edulis*." He sticks out his arm to represent the fleshy part of the mussel's body that protrudes from the shell, and with his hand, he begins to creep along the surface of the lab table. "The mussel uses its foot to shop around for a likely surface, and when it finds one it likes, it cleans it with squirming motions."

We clean surfaces too, he tells me, mainly because our adhesives really need the help. "This table might look smooth, but if you could see its molecular terrain, you'd see hills and valleys—bumps on the surface composed of positive or negative charges. If you wanted a coating of some sort—a sheet of positive charges—to stick, you'd ideally want a surface that had all its negative charges exposed. But if the surface was uneven and some of the negative charges were hidden in valleys, it wouldn't be easy to get a bond. Because our adhesives aren't very talented, we have to spend a lot of time preparing the perfect surface for them. A squirm here and there wouldn't do it for us."

Applying the Primer

After a rather casual cleaning, the mussel presses the tip of its foot down on the surface like a plunger to squeeze the water aside, and then deposits a mucous seal around the edges. Next, the muscles in its foot contract, lifting the ceiling of the plunger and hollowing out a bell-shaped cavity—a vacuum space. Mimicking the vacuum formation, Waite presses his palm perfectly flat on the lab table and then cups it. "Now I'm ready to manufacture a thread and disk, and attach it to the surface with an adhesive."

If only it were that easy for us. Before we can lay down our adhesives, Waite explains, we usually need some sort of primer that will combat water, the bad boy of bonding. Most surface molecules would rather bond with water than just about anything else. And once water grips the surface, the adhesive loses its place (which is why you can usually get a wine label off by soaking the bottle).

A primer is designed to confound water. It occupies the chemical groups on the surface you are painting, in effect hiding the "hooks" that might get caught up in a reaction with water molecules. On glass surfaces (which love water), we prime with silanes, chemicals that imitate the bonds found in the glass itself. While one part of the silane layer occupies the glass, the outward-facing side presents chemical hooks that can bond with the adhesive, or some other polymeric material such as paint.

But even our specialty primers are far from foolproof. If water molecules (in vapor or liquid) manage to enter through a crack or scratch, they'll slip under the adhesive or paint and outcompete the primer, burrowing down to bond with the glass. If we had a talented enough adhesive, suggests the mussel, we wouldn't need primers to achieve good adhesion. And we wouldn't have to worry about our paint blistering or our cars rusting away.

Laying the Adhesive

In the ceiling of the bell-shaped cavity of the mussel's foot are jets that squirt out granules: one- to two-micron-wide balls of liquid proteins that first coalesce, then harden or cure into an adhesive via the cross-linking of tangled strands of protein. In the mussel's case, the cross-linking hooks located on the strands are doubly versatile; they cross-link to one another for *cohesion* (a hanging together of the glue), and they bind to the surface too, in what's called *adhesion*. Conveniently, these hooks are built right into the protein.

The other items needed for the cross-linking reaction—a chemical initiator to kick it off and a catalyst to speed it up—are also right at hand. The initiator for this chemical reaction is oxygen, which comes free for the taking in seawater. A catalyst also comes free, bundled with each mussel protein molecule. After helping to speed the cross-linking, it conveniently becomes a structural part of the glue.

Our adhesives are woefully underequipped by comparison. We have to add not only an initiator to get things going (oxygen isn't enough) and a catalyst to speed things up, but also a separate cross-linking chemical. That's three steps instead of one. Despite all this effort, getting good cohesion and adhesion in one product is still a dream.

Creating the Foamy Plaque

Next, the mussel manufactures the solid-foam disk that anchors the end of the thread. This plaque is made of different proteins that squirt out of jets in the bell-shaped cavity. Once released, they thicken to the consistency of shaving cream and then harden into a solid foam containing air bubbles, like Styrofoam.

"Why a holey substance?" I ask. "Wouldn't a solid mass be sturdier?"

Maybe, says Waite, but sturdiness is not the only thing a mussel needs. Flexibility is also a virtue. A foam will deform more easily than a solid will—allowing it to give a little. This means mussels can perch their plaques on surfaces like pilings or metal stanchions, which expand and contract over the course of a tidal cycle. Whether a mussel is baking in the sun or bathed in cold water, its plaque will give without breaking.

Equally important, a solid foam knows when *not* to give. As Waite explained, "If you notch a solid substance, like glass for instance, and apply force, you'll get a crack propagating 'catastrophically' as the materials scientists like to say. Use a holey material like foam, and the crack will travel only to the first void and then lose steam. It's called a crack-stopping strategy. In wood, the voids are those longitudinal tubes where the sap travels. When you cut a log across the grain you keep hitting them—that's why you stand logs on end in order to split them."

When we make a solid with holes—Styrofoam, for instance—we use what's called a blowing agent to force bubbles into a vat of thickening polymers, or plastic. Unfortunately, the blowing agents of choice are CFCs (chlorofluorocarbons), which, when released to the air, react with the atmosphere and tear up the ozone layer. In light of the hole gnawed in the atmosphere above Antarctica, global leaders have begun to call for bans on the production and use of CFCs. The first phaseout in this country started in 1996, as specified by the Montreal Protocol on Substances that Deplete the Ozone Layer and the 1989 revisions to the Clean Air Act.

With the CFC ban on the horizon, industry was anxious to find a way to make Styrofoam without ozone-depleting chemicals. The military was especially motivated, since it regularly tests explosives against thirty-foot-thick sheets of the stuff. One major consumer, the Picattiny Arsenal in New Jersey, spearheaded research into a CFC-free process.

Its elegant solution answered a question that Waite had been struggling with. "What I couldn't figure out was how the mussel could produce a solid foam without using a blowing agent. When I read about the new gas-free process, I said, of course, this is how mussels must do it! Here we are, toasting the inventors of the new Styrofoam at award ceremonies, not realizing that mussels have been quietly doing the same thing for millions of years."

The old way of making Styrofoam is to pour styrene molecules into organic solvent and wait for them to link into polymer chains thousands of monomers long. As the chain grows, the solution becomes thicker and thicker, eventually turning the consistency of peanut butter and then peanut brittle. Somewhere in between, you blow in a gas to form air spaces—which in technical lingo is called "injecting a gas phase into a liquid phase." No other gas works quite as well as CFCs.

Finally, someone working on the problem thought: Instead of injecting a gas phase, why don't we put a liquid phase into the liquid phase—like oil into water—and have one liquid evaporate while the other solidifies? The big problem was that styrene molecules are just like oil—they hate water and tend to simply settle out in clumps at the bottom of a beaker long before the water evaporates.

The chemists working on this problem should have just taken a break and gone to the biggest salad bar in town. As it turns out, the riddle of keeping an oily liquid suspended in water has a simple solution, one that we benefit from every time we dress our radicchio. Colloidal chemists call it the "salad dressing model."

In prepared dressings, food manufacturers add egg whites to form an emulsion that keeps oil droplets distributed throughout the vinegar so you don't have to keep shaking the bottle. This process works because egg-white proteins are molecules with water-loving heads and fatty, water-fearing tails. To get away from water, the fatty tails all point toward oil droplets, while the water-loving heads stick out into the vinegar. You wind up with separate oil droplets, each surrounded by a skin of egg-white molecules. Carried by these emissaries, the oil droplets stay suspended.

Instead of using egg whites to escort the styrene monomers, the new-Styrofoam researchers used detergent molecules, which are also schizophrenic when it comes to water. Their fatty tails circle around a small group of styrene monomers, forming a "micelle"—a tiny reaction vessel with styrene inside. Literally thousands of these detergent micelles begin to form in the beaker. Inside each one, the styrene monomers begin linking up into a chain. When neighboring micelles collide, the thickening substance from one micelle breaks through its detergent wall and forms a bridge to the growing chain in the next micelle. This happens repeatedly until all the micelles are connected in a giant, solidifying meshwork. Before you know it, the tables have turned, and the water that once surrounded the styrenes is now trapped *inside* their slowly stiffening lattice. As the folks from Picatinny found out, you can pick up the solid lattice, put it on a drying block to wick out all the water, and *voilà!*, you have air inside a solid, sans CFCs!

In technical lingo, this is called a phase inversion. Styrene inside water becomes water inside polystyrene. Waite's theory is that the same phase inversion happens in the mussel's bell jar. The plaque proteins drop into water, and as they cure, the water becomes trapped inside their thickening cross-links. When the water drains out, the mussel has a solid foam plaque containing air bubbles, which is then wrapped in sealant.

I wonder aloud how many other things the lowly mussel had beaten us to, and what we could learn that was new. "We haven't even gotten to the byssus thread," says Waite with a brief smile. He can see me getting hooked on *edulis*, and it pleases him. In his understated way, he is absolutely on fire talking about this bivalve. The lab has long ago emptied out and the lights in the parking lots have flickered on, and neither one of us has budged for hours.

Self-assembling the Thread

The thread is the translucent protein fiber that connects the mussel's soft body to the foamy plaque. "To form the thread," explains Waite, "the entire foot body forms a longitudinal groove, curling in on itself the way some people can curl their tongues. The outer edges of the groove seal and the muscles in the foot balloon out to create a negative space in the groove, a vacuum. Numerous jets along the body of the foot squirt out granules of thread protein, each jet secreting a slightly different variation of protein, custom-mixed to perfection. These proteins are massaged into place by muscles and then left to self-assemble and cross-link.''

When we produce fibers from cross-linked polymer, we, too, use jets to shoot the raw material into a chamber. We do what's called extrusion—a large-diameter screw turns inside the chamber, spiraling the precursor material slowly forward toward a die. The die imposes some sort of ordering or shape as the fiber is extruded, in the same way that a pasta machine makes fettuccine or rigatoni. The difference between us and the mussel is that our fibers are monolithic in character: chains with little or no variety in their subunits, uniform throughout.

The byssus, on the other hand, has a multiple personality. When Waite analyzed the thread, he found that it is made of hundreds of protein molecules, all slightly different in composition. Though their core is collagen protein, like our tendons, each molecule has a portion that is either springy, like natural rubber, or rigid, like natural silk. The proportion of springiness to stiffness depends on where in the thread the protein is located. The molecules at the mollusk end of the thread are springier, while those near the plaque are stiffer, presumably to give the thread the soft-and-hard qualities that it would need in its turbulent home. In testing, Waite found that this customizing of the proteins makes byssus a lot stiffer, tougher, and more elastic than pure collagen would be.

The gradient from the springy top of the thread to the stiff end is not abrupt, however; there is no interface or line drawn between the two. As Paul Calvert had told me, nature loathes fasteners—instead it blends gradients so that the fiber has no single vulnerable point. Waite speculates that such a bifunctional thread would be something we could use for prostheses, or even for robot tendons. The elbow portion of a robot arm could incorporate the rubbery segments, he suggests, while the forearm and upper-arm parts could have stiffer natures. And coating it all, says Waite, could be an *edulis*-inspired sealant that would be even more amazing.

Sealing the Thread

"To me, the transparent sealant that coats and protects the byssus is one of its most exciting features," says Waite. "Byssus is food, after all—it's protein. The only thing that keeps it from being eaten immediately by the voracious microbes in the sea is its sealant."

After the thread and plaque are formed, the whole structure gets coated with yet another set of protein granules that coalesce, spread out evenly, and set to a lacquerlike finish. (The process here is uncannily like the one we use to coat tiny time capsules.) For its finale, the mussel secretes a releasant over everything—a mucuslike substance that allows the newly cast thread and plaque to separate from its mold. Like a curator uncovering a brand-new painting, the mussel removes its foot and the sealed byssus sparkles in the sea light. Although the sealant is itself made of protein, its structure makes it impervious to microbes, at least at first.

"What's neat about the sealant is that it doesn't stay permanently impervious to microbes. The mussel may use its byssus for a few hours or a few days. When it's time to move on, it leaves its byssus behind. In two or three years, the sealant falls apart and the microbes get to feast.

"The reason that excites me," says Waite, "is that we have a lot of consumer products that we use briefly and then throw away." He goes into a lab drawer and pulls out a box of hundreds of pipette tips. He pours them on the slate top and they scatter. "Petrochemically derived plastics like this will virtually last forever in a landfill. Our greatest sin is this overengineering—we may not be able to live forever, but we make darn sure that our waste will."

Waite's idea is to make disposable things that will last only as long as we need them. "We could use natural materials like collagen, silk, rubber, cellulose, or chitin [from crab shells] to produce fibers or containers or whatever, and then seal them with the mussel-type sealant. After two or three years, the sealant breaks down and microbes in the landfill invade the degradable material underneath. Back it goes, into the food chain.

"When you take a natural polymer and coat it with a natural polymer that degrades much more slowly, then you're going toward ideal design that doesn't fly in the face of modern technology. We can still have some throwaway items, but instead of burying or burning them, we can compost them. The degradation can be put off, but not indefinitely the way it is now."

No wonder Waite wants someone to make *edulis* a superhero.

The patents in this one seemingly ordinary animal would support a whole industry. One reason it may have taken innovators so long to look at *edulis* was suggested to me by Randy Lewis, a silk researcher at the University of Wyoming. "Natural materials are difficult to interrogate," he told me. "They're often insoluble proteins, meaning it's tough to get them to separate out. They're usually huge molecules, and until very recently, we haven't had the tools to visualize them. Some of the most interesting are composed of highly repetitive sequences, which, once they are broken into pieces, are like a jigsaw puzzle with only one color—hard to put back together again. As a result, even if funding agencies agree that silk or bioadhesive is an interesting material, they're not certain you'll be able to get to the bottom of it. They usually fund something else that's a surer bet."

Herb Waite has been trying to get to the bottom of a natural material for longer than most. When I ask him how many of the byssus proteins he has left to characterize, he is cagey. "Well, so far we've characterized four proteins called *Mytilus edulis* foot protein or MEFP1 through 4. MEFP1 is the sealant, MEFP2 is the structural molecule in foam, MEFP3 looks like it's present at the foam interface, but that may be a limitation of our technique. I don't know what MEFP4 is yet. We've also got two collagens from the thread, three DOPA-containing proteins [DOPA is 3,4-dihydroxyphenylalanine], and one enzyme. I have to do another DOPA-containing protein and as many as ten minor proteins and an enzyme." Suddenly he stops counting and waves it all away. "I don't really concentrate on how many I have left. It's like climbing a mountain—you don't want to look up and see how far you have to go; it doesn't help. The only thing that helps is to put one foot in front of the other.

"So to speak." And with that, he smiles a very dry Herb Waite smile.

In the meantime, industry has heard about this universal superglue, and companies like Allied Signal are hovering over Waite's work. What intrigues them is the fact that mussel glue will stick to just about anything, probably because of its elegant bifunctional chemistry that cross-links internally while also coupling to a surface.

Once Waite had described the chemistry involved in the cross-linking, Allied Signal cloned what it thought was the gene for the adhesive protein and got *E. coli* to start producing it. Waite also told them that the chemistry depended on a catalyst that cross-links the protein—it converts tyrosine residues into DOPA residues, and then, along with oxygen, they turn into orthoquinones, which are the basis for cross-linking. Though he knew what the catalyst did, Waite still wasn't sure what it looked like. Instead of waiting for Waite to climb that mountain, Allied Signal scientists simply used a common, off-the-shelf catalyst—one that is extracted from mushrooms. "They missed the whole point," says Waite. "The mussel's catalyst is specially constructed to first help with the cross-linking and then to become a structural part of the glue. That's why it's packaged in a one-to-one ratio with the protein. You can't use a nonstructural catalyst and hope to get away with it. You're ignoring the crux of the puzzle."

Sure enough, after years of cloning effort, Allied Signal produced an adhesive protein that wouldn't adhere. "It converted DOPA to quinone but it didn't lead to coating or glue. All we got was a brownish flocculent [a woolly mass at the bottom of the beaker]," says Ina Goldberg, who worked on the research. They decided they couldn't wait for the catalyst to be identified fully, so the research folded.

In the meantime, a group in Massachusetts called Collaborative Research is simply chopping up the mussel foot and selling the purified protein as a cell-and-tissue adhesion product called Celltak. It's not a universal glue yet, but it does work well to coat petri dishes and entice cells to settle down and grow outward in a nice sheet. Word has it that Collaborative Research is about to start marketing a product similar to Celltak that is derived from recombinant DNA. It will sell the plates itself, precoated. In the meantime, a company in Chile is chopping up large cholga mussels—they can be as large as a shoe—and separating out the protein to sell as a petri dish coating.

Using the raw precursors in the foot is one thing, but doing what the mussel does with those precursors is another. No one has yet duplicated the process by which the mussel builds its fiber, its plaque, its adhesive, or its sealant. Waite thinks we may have better luck, in the short term, looking at yet *another* of the mussel's many talents. It seems that the same adhesive protein that binds so adeptly to metals in rocks or on stanchions also clamps on to heavy metals that the mussel ingests in its food. In this way, the mussel stores the toxins in its byssus rather than in its body, and when it moves to greener pastures, it jettisons the byssus and leaves the heavy metals behind.

The U.S. Environmental Protection Agency (EPA) is interested in the record of metal accumulation that's left in that cast-off byssus. In its program called Mussel Watch, the EPA harvests byssus leftovers in the Chesapeake Bay, and analyzes them over a period of time to see if metal residues in the bay are trending up or down. Waite can envision cloning the gene for that protein (so we can make massive quantities), and then using it as a screen in a filtering system. The protein filters could be installed on ships, dragged for a time, and then analyzed for metal residues.

"It's only one of the many practical inventions that could come from the mussel's repertoire," says Waite. "As we perfect our technologies, I'm sure we'll run across other processes and designs that *edulis* has already worked out. The adhesive is only one patent among many."

And *edulis*, of course, is just one bivalve among many, one invertebrate in the ocean among many. Suddenly I wish it were Herb Waite we were cloning, instead of just proteins.

For the reasons that Randy Lewis listed, there are not many like Waite who have decided to tackle natural materials. Although many engineers admit that there's merit to this inquiry, the obstacles make it a long-term, hair-pulling endeavor. "You have to be sure the material is really worth it," says Lewis. One material that has won over many researchers, including Lewis, is a 380-million-year-old fiber with a twenty-first-century future. Spider silk, says the University of Washington's Christopher Viney, is the stuff that dreams are made of.

Along Came a Spider

It's a steamy 80 degrees F. in Christopher Viney's Seattle lab, in deference to Tiny, a six-inch-long golden orb weaver spider (*Nephila clavipes*), who is now flipped on her back, dining on crickets while being silked. A gossamer thread issues from her enormous abdomen at a steady clip, wound by a motor onto a revolving spindle. In this session alone, Tiny will donate about one hundred feet of "dragline," a specialty silk designed for rappelling from drop-offs and framing the spokes and perimeter of her web.

Dragline is only one of six silks that this eight-legged factory can produce, each one mixed in its own gland, extruded through its own spinneret, and endowed with its own chemical and physical properties, all of which the spider needs to survive. As the late arachnologist Theodore H. Savory once remarked, "Silk is the warp and woof of the spider's life."

Many spiders begin their lives as eggs swaddled in silk and take their first trip via a thin strand that catches on air currents and "balloons" them to new, distant homes. When hunger strikes, some spiders spin a nearly invisible snare, while others spin dense sticky sheets that snag insects the way flypaper does. Still others dispense with web spinning altogether, simply extruding a single silken strand with a sticky ball attached. "The ball is hurled, gaucho-style, at insects flying by, which are then lassoed in and calf-roped," writes entomologist May R. Berenbaum in her book *Bugs in the System*. Silk also figures prominently in the sex lives of spiders. In courtship, silk may be laced with pheromones (sex attractants), like a handkerchief sprayed with cologne. Once the wooing has worked, males may spin more silk to immobilize the female (who is just as likely to eat her suitor as to mate with him). Still not wanting to get too close, he deposits his sperm into a special little package of webbing, which he inserts into the female. Even in death, writes Berenbaum, spiders' lives are tied up in silk. Certain species of spiders are known to wrap the remains of a dead compatriot in specially woven shrouds.

Lately, this mysterious material has also become central to the lives of a small cadre of materials scientists. As Christopher Viney drops another cricket Tiny's way, he seems more surprised than I that his career has come to this. "I'm a metallurgist!" he says, feigning defensiveness. "Really! I'm a licensed physicist! I haven't taken a biology class since high school!" I begin to pick up some of the paraphernalia festooning his room—a rubber spider, macramé spiderwebs, a can of slug chowder ("Please don't add salt," the label cautions), biology journals, an article that refers to him as the Spider Man. "OK." He throws open his large hands and shrugs. "So I went astray."

"Astray" began in high school in South Africa when Viney had a biology teacher who was also a museum curator. "He veered wildly off the syllabus, regaling us with stories about cracking the DNA code and other exciting developments going on at the moment in science. His enthusiasm was absolutely infectious. As a result, when I applied to Cambridge, I actually did better on my entrance exams in biology than I did in physics and chemistry, which was what I wanted to go into. I eventually wound up studying metallurgy in the Natural Sciences program, which was the most interdisciplinary option available. I didn't learn a thing about welding, but I did learn about atoms and molecules."

One of the most important classes Viney took was an elective that taught him a skill he would later use while surfing between disciplines: crystallography. Crystallography is the study of how organic and inorganic materials, under certain conditions, assume very ordered shapes and structures called crystals. The atoms in a crystal line up in predictable spacings and stay that way, giving you something like three-dimensional wallpaper, with a pattern that repeats itself in all directions. A liquid has a much more random arrangement of molecules. There is no pattern to help you describe or predict exactly where the molecules are.

In between the order of a crystal and the disorder of a liquid is a material called a liquid crystal, which has some qualities of both. It's a liquid with its molecules arranged in orientational but not positional order; that is, the molecules are all aligned in some dimension—they're facing the same way—but they aren't positioned in a predictable pattern. Though Viney didn't know it at the time, his early fascination with these semi-ordered crystals would lead him directly into Tiny's web.

"Actually, it all started one Saturday night while I was on the couch reading dirty physics magazines," he laughs. "I came across an article by Robert Greenler [physics professor, University of Wisconsin-Milwaukee, and president of the American Optical Society] on why you can see rainbows in spiderwebs at dawn and dusk. It combined optics, which I love, with silk, which I knew very little about. As it turns out, no one else did either. We'd been cultivating silkworm silk for four thousand years, but when Greenler needed the refractive index (a very common measurement) for spider silk, he had to guess at it.

"This made me curious about the refractive index of spider silk. I did a test and realized that it was very high. Usually, a high refractive index points to some sort of crystallinity, and that's just what we found in spider silk—small crystallites embedded in a rubbery matrix of organic polymer. Somehow the spider had learned to manufacture a composite [two types of material in one], three hundred eighty million years before we decided composites would be all the rage!"

As a metallurgist, Viney knew that this unusual structure must impart an equally unusual function. Sure enough, the stellar properties of spider silk are enough to make materials scientists suspect typos. Compared ounce to ounce with steel, dragline silk is five times stronger, and compared to Kevlar (found in bulletproof vests), it's much tougher—able to absorb five times the impact force without breaking. Besides being very strong and very tough, it also manages to be highly elastic, a hat trick that is rare in any one material. If you suspend increasingly heavy weights from a steel wire and a silk fiber of the same diameter, their breaking point is about the same. But if a gale force wind blows, the strand of silk (five times lighter in weight) will do something the steel never could—it will stretch 40 percent longer than its original length and bounce back good as new. Up against our stretchiest nylon, spider silk bungees 30 percent farther.

This energy-absorbing elasticity comes in handy when moths and other "meals on wings" come hurtling into the web at top speed. Instead of breaking, the gossamer strands stretch, giving off most of their impact energy as heat. Fully spent, the web recoils so gently that it doesn't trampoline the moth back out. "None of our metals or high-strength fibers can come even close to this combination of strength and energy-absorbing elasticity," says Viney. According to Science News reporter Richard Lipkin, in a January 21, 1995, article, spider silk is so strong and resilient that on the human scale, a web resembling a fishing net could catch a passenger plane in flight!

Another characteristic in silk's favor is its unusually low glass-transition temperature. This simply means that silk has to get very, very cold before it becomes brittle enough to break easily. In the frigid temperatures that parachutes encounter, for instance, spider silk would make ideal lightweight lines. Other uses for a fiber as strong as spider silk would be bulletproof fabrics, cable for suspension bridges, artificial ligaments, and sutures, to name just a few. The question is, how would we go about packing so much function into such a small package?

Spider silk begins as a pool of raw liquid protein sloshing around in a gland that Viney says looks like "the business end of a bagpipe." The raw silk (a liquid protein) travels from the gland to a narrow duct before being squeezed through one of the six spinnerets—minute groups of nozzles at the spider's back end. The miracle is that what goes into the spinneret as soluble liquid protein (easily dissolved in water), somehow emerges as an insoluble, nearly waterproof, highly ordered fiber. "It's enough to make a fiber manufacturer very jealous."

Viney guessed that the raw silk somehow went through a liquid crystal phase just before squeezing through the spinneret. This would align the molecules and give them a jump on their ordering. To be capable of achieving the liquid crystal state, Viney figured, the sub-units—proteins—would have to be "anisotropic" in structure. "An anisotropic substance is one that has a definite directional order," says Viney. "The uncooked strands of spaghetti in a box are aniso-tropic. They look different depending on whether you are viewing them end on or from the side. The opposite of anisotropic would be an isotropic tangle of cooked spaghetti, which looks the same in all directions. Although most people thought soluble spider protein was isotropic, I was expecting to see anisotropic rods of some sort."

One of the best tests for anisotropy would be to look at the raw silk under a polarizing light microscope, an instrument invented over one hundred years ago, which fewer and fewer people know how to use. Not only did Viney know the instrument, he had become some-what of an expert, even writing a modern-day manual on its use. "The polarizing light microscope uses the same principle as polarized sunglasses. Only instead of one filter, it has two—one cuts out every-thing except light vibrating vertically, while the other cuts out every-thing except light vibrating horizontally. For most objects, this accounts for all the light passing through, so you see only darkness in the scope. An anisotropic material, however, plays with the po-larization state of light." When Viney looked at liquid spider silk, especially at the edges of a slide where it was drying, he clearly saw light coming through the filters, a sure sign of anisotropy. "In fact, according to the patterns we saw under the scope, it looked to be a rod that was thirty times longer than it was wide."

To check his hunch, Viney consulted the protein sequencing data published by Randy Lewis of the University of Wyoming and Dave Kaplan of the U.S. Army, only to meet with more frustration.

Pop Beads and Slinkys
The amino acid sequences of raw liquid silk didn't seem to corre-spond to any protein that would fold up into a rod. In fact, the repetitious sequences pointed to a protein that, while it was in the gland, was most likely to be tangled and globular, "like a ball of wool the cat's gotten hold of." The water-fearing amino acids in the chain were probably hiding in the middle of the ball while the water-loving amino acids hung on the periphery. This arrangement wouldn't change until the ball was physically sheared by the spinneret.

In a way, this made sense. Globular molecules floating in water would be a good way to store the protein in the gland. When the spider twisted and scurried through its days, the globules would simply roll with the punches, and the spider didn't have to fear "becoming constipated with its own silk" if the liquid protein somehow sheared into fiber form. But if there were *only* globular molecules, thought Viney, why was the polarizing light microscope showing undeniable evidence of rodlike structures?

"The mystery unraveled for me when I attended a lecture by one of my colleagues in the bioengineering department," he says. The speaker was talking about actin, a protein that self-assembles to help form our muscles. Actin is essentially a globular protein, but the balls hook up to one another—like the baubles in a kid's pop-bead necklace—to form a chain. As Viney looked at the cartoon graphic, something breached and leaped from his subconscious.

"There was my rod!" he said.

Viney turns on his computer and we look at cartoon depictions of his evolving theory of spider silk formation. He now hypothesizes that the raw liquid silk leaves the gland and travels through a thin duct just before entering the spinneret. As it squeezes through the duct, water is wrung out of the protein and calcium is added. (Calcium is what allows actin globules to hook up, so Viney thinks it may also be at work here.) The globules hook up in a pop-bead necklace, making the solution one thousand times less viscous, because the rodlike assemblies can now slide past one another. It's analogous to putting lanes of traffic on a highway sliding past one another, versus the mess that is a laneless, lawless Manhattan jam.

Connected, aligned molecules are not only easier to push through the spinneret, they are also more susceptible to the shearing action that turns liquid protein into fiber. Because the globes are unable to roll out of the way, the squeeze through the spinneret disrupts the water-loving residues on their periphery, exposing their water-fearing parts.

"These hydrophobic parts go 'ARRRGG!' and cluster together as tightly as they can," says Viney. They assume a zigzag shape, folded accordion-style into pleats. One pleated sheet stacks on top of another, as close as they can get to lock out the water. The water-loving portions of the proteins remain loose and curly at the edges, forming the springy matrix that the accordion crystal parts are embedded in.

Viney's model has a pleasing simplicity and completeness: The globular proteins line up into a pop-bead necklace, which squeezes through the spinneret to become a silk fiber. The final product is partly flexible and partly rigid, like a reinforced Slinky. The amorphous part gives, but the stiff crystalline domains don't give. When the fiber becomes notched, a crack or tear gets interrupted by the crystalline regions and can't propagate. The model also explains why the material goes from being a soluble liquid to an insoluble fiber. Once the water-fearing portions of the proteins crowd together, they resist water, ensuring that the silk won't fall apart.

Nevertheless, it's only a model, and some, like silk researcher Randy Lewis, don't agree with it. Lewis feels he has evidence that there are actually two proteins rather than just one that make up spider silk. "In the two-protein hypothesis, Viney's pop-bead model doesn't make sense," says Lewis. But other researchers, including Viney, are still not convinced of the existence of two proteins. While the jury is out and the debate is lively, all the investigators in spider silk research encourage one another to keep theorizing. When you think about what it could mean in terms of sustainable fiber manufacture, this research, tough as it is, is definitely worth it.

Consider: The only thing we have that comes close to silk in quality is polyaramid Kevlar, a fiber so tough it can stop bullets. But to make Kevlar, we pour petroleum-derived molecules into a pressurized vat of concentrated sulfuric acid and boil it at several hundred degrees Fahrenheit in order to force it into a liquid crystal form. We then subject it to high pressures to force the fibers into alignment as we draw them out. The energy input is extreme and the toxic byproducts are odious.

The spider manages to make an equally strong and much tougher fiber at body temperature, without high pressures, heat, or corrosive acids. Best of all, says Viney, spiders don't have to drill offshore for oil to produce the silk. They take flies and crickets at one end and process a high-tech material at the other end.

If we could learn to do what the spider does, we could take a soluble raw material that is infinitely renewable and make a superstrong water-insoluble fiber with negligible energy inputs and no toxic outputs. We could apply that processing strategy to any number of fiber precursors. Imagine what it would do to our fiber industry, which is now heavily dependent on petroleum, both for raw material and processing! To break that dependency, says Viney, we have to become spider's apprentices. "If we want to manufacture something that's at least as good as spider silk, we have to duplicate the processing regime that spiders use. We have to mix up a batch of precursor and duplicate the physical journey from the glands out to the spinnerets. It's that journey that helps impart a certain microstructure to the fibers.

"When we scale this journey up for industrial use, we have to be able to give the manufacturing crew exact specifications: what concentration of protein they should use, how big the rods in the liquid crystal should be, how much calcium they will need, how much water they should squeeze out, and how fast they should spin out the fibers to obtain a silk with desired properties. By tweaking any one of those variables, we may be able to customize the silk for different uses. For me, the processing is the really intriguing part of this story."

While Viney works on scaling up the process, there are other scientists examining the protein precursors that will make all of this possible. Silk is after all a biological material—a protein that self-assembles, under a gentle shear force, into a fiber. The protein hunters I visited are deep in the gland of the orb weaver, hoping to characterize the source of silk and find a way to produce it without Tiny's help.

Silk Maneuvers

When I first laid eyes on David L. Kaplan, it was in a photo, one of those rare shots that captures a person's essence. He was standing behind a glass case, peering not at the camera but at a six-inch-long orb weaver spider. His eyes absolutely shone with entrancement—like the eyes of a child at the zoo, staring through a window at an animal that is staring back at him.

Kaplan is entranced for nearly fourteen hours a day, arriving at the U.S. Army's Research, Development and Engineering Center in Natick, Massachusetts, long before his employees arrive, and leaving long after they leave. "It's never a dull moment," he tells me. "You learn one thing about nature, and you come up for air with ten more things to pursue. We are right on the verge of so much, so much that has no precedent."

Kaplan is always on the move, usually trailing a person or two who wants to see him. He directs forty-five people in all (those I talked to rave about him) and is responsible for overseeing the technical aspects of every study going on in the biomolecular materials department at Natick. One of his favorite projects is the quest to synthesize a gene for a silklike protein.

The army wants a fiber that protects better than Kevlar, and is willing to look to nature for it. They want it to be ethereally light, yet strong enough to bundle into cables for suspension bridges or for bungees that hook fighter planes as they scream onto aircraft carriers. And yes, they suppose it would be nice if its manufacturing process was environmentally friendly to boot. Kaplan explains how his team plans to deliver.

"We first had to find a way to efficiently remove the liquid protein from spider glands without it turning to silk, and we did, but this gave us only an infinitesimal amount. If we ever hoped to commercially produce this stuff, we knew we would have to take a genetic approach.

"Our first shot out of the barrel was to isolate the full-length natural gene—the native gene. It's a whopping nine to ten kilobases [a kilobase is a thousand DNA subunits strung together] and a nightmare to work with because it's highly repetitive and prone to deletions and recombinations when it's expressed by *E. coli*. We're still doing some work on the wild gene, but we realized we'd increase our chances of producing some kind of silk precursor if we also tried to synthesize a simpler gene on our own.

"The techniques are not there yet to synthesize a DNA fragment as long as the natural one. Instead, we could only hope to synthesize a small DNA fragment [with an oligio machine], multiply it, and glue those fragments together with ligases [enzymes that help the fragments to combine]. Deciding *which part* of the long genetic sequence to choose for synthesizing—that was a very educated guess. You learn what you can, and ultimately you go with your gut, and that's where science turns to art. We were also guessing when it came to the different ways we glued the DNA fragments together. We matched the codon preferences for *E. coli* as best we could and then we started the arduous part—trying to coax *E. coli* to accept our homemade DNA and make the protein for us.

"To make a long story short, it worked. *E. coli* expressed the protein, giving us something to test and learn from. We want to see what properties it has, and then figure out why: What is it about this amino acid sequence that might have given rise to these properties? What we're searching for is some correlation between the structure of the protein—how its amino acids are arranged—and its function. We're after some rules of thumb. Ultimately, we'd like to be able to tell a fiber designer, if you want resilience, try this repeating sequence of amino acids followed by that repeating sequence. Slowly but surely, we're building an information infrastructure—a knowledge base that will allow us to make materials the way nature does. What we learn from spiders will be helpful for any polymer processing. We're not the only ones by any means. Randy Lewis is working with the wild gene out your way," he tells me.

In the windy town of Laramie, Wyoming, Randolph V. Lewis has the sequences to what he believes are two proteins at work in the spider's gland. His team at the University of Wyoming used genetic engineering "probe" techniques like those used on abalone shell to isolate portions of the two genes that code for the thread-producing proteins. They then inserted these gene fragments (each representing only about one third of the real genes) into *E. coli* and successfully expressed proteins. "We even processed them into fibers, but they didn't match the qualities of dragline silk. Our truncated genes were obviously missing something important." Now Lewis's team, like Kaplan's, is working to synthesize a gene that may come closer to the qualities materials scientists are looking for.

Lewis is applying for grants that will allow him to analyze different kinds of silk from different kinds of spiders, hoping to learn more about the structure-function relationships Kaplan talked about. He, too, is trying to come up with the ultimate cookbook that will allow fiber manufacturers to look up the properties they want in a silk protein, then find the amino acid recipe for those properties. Want a better fiber? Start with a better protein, and template your fibers to taste.

Lewis's foray into different spiders and different silks makes me wonder whether we are studying the best possible models. All of our current knowledge comes from studies of only two kinds of threads spun by fewer than fifteen species of orb weavers, a subset that makes up only one third of all thirty thousand described spider species. Is an even better prototype waiting out there somewhere?

Back in Seattle, I pose this question to Christopher Viney, whose normally cheerful, mischievous face clouds. He thinks carefully. As in all of biology, model systems are chosen because they are easy to work with, he explains, and to some extent because other people have already set the track for you. But yes, there probably is a stronger, tougher, stiffer fiber being produced right this minute by a spider that we know nothing about. A spider whose habitat may be going up in smoke.

"And do I feel an urgency to learn what I can before these models go extinct?" he asks. He looks around his office, out the window, and after a while, back at me, as serious as I've yet seen him. "Well," he says, in that way the British have of retiring from a topic. "I suppose it won't hurt to have my metallurgy wait a few more years." He stands and I stand, and for the first time all day, we check the clock.

Horns for the Rhino's Dilemma

With species like the rhinoceros, the countdown to extinction is not just speculation—it's an ongoing spectacle. There are only 2,300 black rhinos in all of Africa, down from 65,000 as recently as 1970. Zimbabwe's wild population, thought to be 1,400 in mid-1991, is down to a shocking 250 animals. Asian rhinos are faring no better. The Sumatran rhino population has been cut in half in the last ten years, with numbers now totaling fewer than 600.

The reason rhinos are in decline is because of the five to ten pounds of protein that shapes itself into a horn (or two) protruding unicorn-style from the rhino's head. Poachers risk being shot on sight to kill a rhino, but if they can get the horn and get away, they earn an amount equal to a year's wages. The horn lords reap the real money, though—they sell the horns on the black market for tens of thousands of dollars each. Half of the horns used to go to the Middle East, where they were crafted into dagger handles that Yemeni men strapped on during their ritual initiation into manhood. A single dagger may have cost $30,000, but the status conferred on the man who wore it was deemed worth the price. These days, most horns find their way into Eastern medicines. Rhino horn powder is thought to cure stomachaches, skin blemishes, a lackluster libido, and even, purportedly, a lousy singing voice.

Although legal sales have diminished since the 1977 international ban on sale of rhino horn (CITES), rhinos continue to be killed and the horns bled off slowly to the black market. Poaching has gotten so bad in Namibia that officials began a dehorning program, sawing off the trophy as a way to spare the animals' lives. Perversely, the slaughter continues, this time with dehorned rhinos. "What we now think is that the horn lords are hoping to see the rhinos go extinct, which will increase the value of their stockpiled horns," says Joe Daniel, a rhino researcher at Old Dominion University in Virginia.

I traveled to Old Dominion because I had heard that Daniel, a zoologist by training, had teamed up with a metallurgist named Ann Van Orden, and they had a plan to help stop the slaughter. It would be biomimicry at its best.

What the world needs, say Daniel and Van Orden, is a facsimile rhinoceros horn that is inexpensive to make. "Flooding the market with this horn, identifying it as a facsimile and hoping to get other cultures to accept it—that may be our only option. Or rather, the rhino's only option. If we work it out so that horn lords are able to make a profit in volume selling, they may decide it is no longer worth their while to risk the poaching."

History bears them out. Whenever we have given people convincing substitutes for a coveted material, it has helped to conserve the original. Rubber trees were not so heavily tapped, for instance, nor pearls so voraciously fished, after artificial substitutes became available. The key is to offer a duplicate material that is almost as lustrous, almost as rubbery as the real thing. The lower price speaks for itself, and in the process, native organisms are freed from our hungry grasp.

But rhino horn is an especially tough case for mimickers. Haloed as it is with magical and medicinal qualities, consumers are loath to accept any substitutes. When I asked what the horn is made of, Ann Van Orden drummed the table with her fingernails. "It's keratin—the same tough, fibrous protein that's in your fingernails and your hair. There's absolutely no proof that rhino horn can do what it's touted to do, no more than your ground-up fingernails could. It's not the keratin itself, however, but the unique way that it's structured that gives rhino horn its coveted strength and luster. If we could induce keratin to self-assemble into that structure, we'd have the viable substitute we need."

The two collaborators who hope to pull this off met serendipitously when Van Orden's husband, a physicist at Old Dominion, came to Daniel's brown-bag seminar on infrasound and rhinos. When Daniel mentioned he needed some help preparing samples of rhino horn for the microscope, Ann's husband suggested her for the job. Van Orden picks up the story: "I was working at Langley Research Center at the time studying corrosion, and needless to say, rhinos were not in my annual work plan. So I code-named my folder

Rufus (the name of the bull at Virginia Zoological Park who had donated a broken-off piece of his horn) and kept it to myself."

At lunch, Van Orden lowered an unpolished chunk of Rufus's horn into my palm. "Be careful," she joked. "What you're holding is worth about ten thousand dollars." At the fractured edge, I could see fibers called spicules sticking out. They were tipped with points at each end, like porcupine quills. "This is an ingenious design. As one of those spicules tapers to a point, it makes room for another tapering spicule to take off. In this way, the fibers are interdigitated, and that's why you see a zigzag break when the horn breaks—some points are sticking out, some holes are left behind."

But where's the hair? I ask her. In almost every book I'd read (and one that I'd written), rhino horn had been described as bundles of hair tightly packed together. She smiled. "I know, but that's not what I saw when I sectioned it." Chances are, no biology department had ever sectioned and prepared the horn for microscopy in exactly the way Van Orden did. She treated it as she might a piece of metal that had corroded. She sawed a cross-sectional slice and sanded it, starting with 300-grit sandpaper, working up to 1,200-grit, and finally polishing it to a scratch-free finish with a diamond slurry and an alumina polish. She then examined it under a polarizing light microscope (the type Viney used on spider silk) as if it were a piece of metal. A color photo of her cross section was hanging in her office with a blue ribbon on it, having won first prize in Polaroid's science photography contest.

The horn was indeed beautiful in cross section. It was as if someone had taken a bundle of solid, copper-colored quills and cut across them, leaving a landscape of what looked like cells. The softer centers of the spicules had given way under the sander, leaving little concave depressions in the middle of each cell. As Van Orden explained, the concave depression is the central core of the spicule; that core is a fiber that grows from a follicle at the base of the rhino's horn. Around this core, keratin-producing cells—now dead, flattened, and cornified like skin cells—lay in concentric fashion, looking like growth rings on a tree trunk. They produce what amounts to a hard, multilayered keratin sleeve around each fiber. Around the outside of this sleeve, there's another kind of keratin, also fibrous, that serves as the matrix or mortar between the spicules. Despite what all the textbooks said, the horn was not hair at all; it was a composite made of two forms of keratin in the same material.

When Van Orden looked at the magnified horn slice with her materials scientist eyes, she immediately recognized the pattern: "It looked just like the graphite–fiber-reinforced composite we use for the skin of the Stealth bomber! You take graphite fibers, which are very stiff and inflexible—they break before they bend—and encase them in a resin which is flexible—it bends before it breaks. You wind up with something rigid but very tough to break. That's why composites are so wonderful; they add up to something more than the sum of their parts."

Besides the Stealth bomber, graphite-reinforced composites are also used in the masts on America's Cup sailboats, the bodies of Formula One race cars, high-end guitars, tennis rackets, and Boeing's new lightweight 777 airplane, which will fly farther and faster on less fuel. "In engineering this composite, it seems we've coevolved," says Van Orden. "We've invented something nature has already been using for sixty million years."

Next she showed me a picture of another kind of composite— silicon carbide fibers embedded in aluminum oxide (a ceramic matrix)—and pointed out the differences. "We make this composite by laying fibers down by hand and then putting a block of ceramic on top. We combine the two under pressure and heat, so the ceramic resolidifies in and around the fibers. We have to be careful to keep the fibers a certain distance apart, though, because if they diffuse into one another under those extreme conditions, the resulting composite is not as resistant to breakage."

Because our heat-and-beat processing doesn't lend itself to that level of control, we can't optimize the way nature can. The rhino's horn, which self-assembles from within, has densely packed spicules that are carefully spaced and not touching. This better "packing density" makes for a tougher horn. Matrix and fiber are also chemically similar, and consequently are able to bond well at the interfaces.

Another difference between our composite and nature's is in the shape of the fibers. In cross section, the synthetic fibers in the carbon-graphite composite are uniformly round, while the rhino fibers vary in size and shape. What *does* remain uniformly thick throughout is the mortar—the keratin matrix.

"Again, this makes great sense from a materials science point of view," says Van Orden. "It may be that a certain thickness of matrix has to be there as a buffer against insults. If the fibers were abutted together without a buffer, and you broke one, you could break them all." This way, the keratin matrix acts like the mortar in abalone shell; it interrupts cracks coming from the side, and the stress is redistributed, giving the horn torsional strength.

But what does this design buy the rhino? Daniel tells me that cow rhinos wave, joust, and stab with their horns to protect their calves from attack. Bulls use it to drive off interlopers in territorial disputes, and all rhinos use it for digging in the ground. To be able to do all this, rhino horn has to possess strength when pushed from the tip as well as from the side. This head-on strength is called compressive, and building a spicule shaped like a porcupine quill is a great way to achieve it. Van Orden shows me a close-up of a horn fracture, in which the tips of the quill-like spicules are all bent. "Here's where the compressive strength comes in—instead of being flat topped, and taking all the pushing energy directly, it's tipped with a point. The point simply bends or breaks, but doesn't transmit the load all the way down. There's a lesson we should apply right now to our composites, which are rarely blessed with both compressional and torsional strength."

What has Daniel and Van Orden really excited is a third trait of rhino horn that our materials don't possess—the ability to heal. The evidence of self-healing was hiding in the beautiful Polaroid picture. "If you look closely, you'll see a crack that has infilled with polymer, essentially healing over," says Daniel. "But as a biologist, I considered this impossible, because as far as we know, there are no living cells in the horn—only dead tissue. Or so we thought. The idea that there might be something alive in the horn opened the possibility that we could take a sample of those cells and try growing, through tissue culture techniques, a horn in vitro."

In search of living cells, Daniel went to an exotic-animal breeding facility in Texas to perform a needle biopsy on a rhino. "The veterinarian agreed to call me when he was going to do a checkup, because they don't like to anesthetize the animals very often. I flew down there, took the horn sample, and then placed it into a liquid growing medium. If keratinocytes [cells] had been present, they would have grown out into that medium. Unfortunately, that didn't happen. Now I'm waiting for another rhino to need its shots so I can fly down for another biopsy."

In the meantime, Daniels and Van Orden are investigating other options for making a horn facsimile. "Say the healing isn't caused by living cells," says Daniels. "Say instead that the material for the infilling is cannibalized from somewhere nearby. Say a portion of the horn depolymerizes [breaks down into its building blocks], flows to the crack, then repolymerizes to fill in the gap. That got us thinking—maybe we could do something similar. Maybe we could depolymerize rhino keratin and induce it to reassemble around a core of rhino hair."

Daniel's idea is to practice on something like horsehair. Horses have two types of hair—the rough tail hair that is used for violin bows, and the softer hair of the coat. First he would depolymerize the coat hair into a liquid, then lay the tail hairs side by side in the liquid solution and put it under pressure. The keratin would, it's hoped, polymerize around the nucleus of the larger fibers, forming connectors that gather all the hairs together.

The same sort of technique is already being done with bone, says Van Orden. "A dental surgeon can take bone and treat it so that just the hydroxyapatite is left. To build up the jawbone foundation beneath an implant, for instance, they'll cut open a patient's jaw and put in this hydroxyapatite. When the person's bone cells come in contact with this hydroxyapatite, they say, 'Hey, we forgot to calcify this!' and they'll make new bone in that spot. We're hoping that our liquefied horsehair cells might see the tail hairs and say, 'Oh, hair tissue. We forgot to gather all this together.' If it works with horsehair, we'll do the same thing with rhino keratin—we'll provide the hair, the keratin, and the right conditions and say, 'Structure yourself!' "

Although it seems like a blue-sky idea to grow rhino horns from scratch this way, it may not be. "Heck," says Van Orden with her trademark enthusiasm. "If you told us thirty years ago we'd be putting graphite fibers in a resin matrix, it would have seemed far-fetched. Today we're playing doubles tennis with composites like this."

Thirty years ago, there were many, many more rhinos than there are today. No matter how far-fetched or whatever else may come from this research in terms of composite innovations, any attempts to stop rhino slaughter will be well worth the effort. In fact, it's one of the best uses of biomimicry I can imagine. This time, we're learning to imitate an animal not to save ourselves (directly) but to save another species from "an end to birth." It's biomimicry come full circle, a glimpse of the good we could do with this new science if we choose to.

That sets me thinking about which agencies, or which foundations, have had the foresight to sponsor this type of research. When

I ask Daniel and Van Orden, they lock eyes across the table and simultaneously make the hand sign for zilch. Their rhino horn work has received no official funding. For now, their quest is a labor of love and conscience, a pro bono for Rufus and the thinning herds whose horns alone, strong as they are, cannot protect them now.

CHAPTER 5

HOW WILL WE HEAL OURSELVES?

EXPERTS
IN OUR MIDST:
FINDING
CURES
LIKE A CHIMP

Nature is the supreme chemist. With all due respect to the brilliance of chemists, I don't think a chemist could dream up a molecule like Taxol. [Taxol, a promising new cancer drug, is found in the bark of the Pacific yew tree (Taxus brevifolia) *in the Pacific Northwest.]*

—GORDON CRAGG, chief of the natural products branch,
National Cancer Laboratory, Frederick, Maryland

What matters is that we swallow our hubris and start acknowledging that animals have many things to teach us.

—RICHARD WRANGHAM, MICHAEL HUFFMAN, KAREN STRIER,
and ELOY RODRIGUEZ, zoopharmacognosy pioneers

Kenneth Glander's office at the Duke University Primate Center in Durham, North Carolina, has a perfectly round, high-dome ceiling that feels like it might be thatched, and that an African village might be right outside. Instead, the back door leads to the Carolina piney woods, a humidor of coniferous spice and old, fuming leaves. In the morning drizzle, Glander tilts back his head, catching sparkling drops on the drooping handles of his waxed mustache. In the uppermost canopy of trees, small balls of fur curl up against the rain: five hundred lemurs, some of the most painfully endangered primates in the world. Aye-ayes, sifakas, and other prosimians are being bred in this arboreal ark in the event that they become completely zeroed out in the wild. Part of Glander's mission as ark director is to see to it that they keep themselves healthy.

But these woods are half a world away and vegetatively different from the lemurs' home in Madagascar. "It took me five years to convince people you could let these animals roam in these woods without fear of them poisoning themselves on our mushrooms. Even though people die from eating mushrooms all the time, I had a hunch these primates were smarter than that."

Primates *are* smarter than that, and so are elephants, bears, birds, and even insects. Wild things live in a chemically charged world, and their goal in life is to pick their way through the maze of poisons and find a packet of energy or perhaps a dose of curative. We humans were once as omnivorous as they, able to pick and choose between the good, the bad, and the bitter.

Today, we are beginning to return to wild places to search for new drugs and new crops (or wild genes to add spunk to our old standbys). Given our domesticated and dulled senses of taste and smell, however, we now screen the forest for promising plants in a time-consuming way. Instead of innately sensing the best, we collect it all and painstakingly sort through it. Given the rapid acceleration of plant extinctions, we no longer have time for this buckshot approach.

There are more than four hundred thousand plants and as many unique chemicals that we have yet to explore as possible medicines or foods. Before they're all gone, say the biomimics practicing "biorational" drug and crop discovery, we need to consult the talented taste buds of wild connoisseurs and fur-covered pharmacists. They have, after all, been "native to this place" for millions of years longer than even our most astute agronomists or medicine men. They know what to eat and what to avoid, what will make them sick, delay the birth of an offspring, give them energy, or arrest a case of diarrhea. They are the experts we have been too arrogant to consult. Now, in this era of massive loss and little time for screening, we are beginning to tap them on their furry, scaled, feathered, and exoskeletoned shoulders and ask, "What's that you're eating?"

CHEMICAL WARFARE, PASSIONFLOWER STYLE

In order to appreciate the gustatory talent these wild experts possess, it helps to focus your mind on a fine hallucination. Think of yourself as a plant, rooted in place, unable to switch your tail or twitch your flanks. You are the succulent object of desire for countless microbes, insects, and animals that can't photosynthesize their own food. You may parry their attacks with leathery leaves, thorns, or perhaps burrowing nettles, but your warfare of choice is chemical.

The stew of so-called "secondary compounds" that you, the plant, produce are what gives the green world its flavors, fragrances, spices, medicines, and poisons. It is with these chemicals that you bite back—burning, revolting, intoxicating, or even killing those that dare to eat too much of you.

Now focus on another hallucination. Imagine yourself as a wild herbivore confronted with a jungle full of defensive plants, each one doing its best to get you to keep your big square teeth to yourself. It would make a good computer game, actually. Here are the rules: Armed with only your senses, your powers of observation, and your memory, you have to gather your own food. Before moving on to the next level of play—surviving long enough to pass on your genes—you have to garner just the right amounts of vitamins, essential amino acids, proteins, and other nutrients to survive.

It may look like the Garden of Eden out there, but nature's menu is a minefield. Even if a food doesn't kill you outright, its secondary compounds can rob you of nourishment. The lineup of plant poisons includes alkaloids, phenolics, tannins, cyanogenic glycosides, and terpenoids, all possessing devilish ways to discourage digestion. Alkaloids such as nicotine and morphine, for instance, interfere with your nervous system. Cyanide (a tannin) and cardiac glycosides dive straight into your muscles, wreaking havoc with your heart rhythm. The respiratory inhibitor in passionflower (cyanogenic glycoside) will literally take your breath away. Or, if you'd like, plant hallucinogens will liberate you from your good sense and get the plant off the hook in the process. (As ecologist Paul Ehrlich says, "If a deer nibbles a hallucinogenic plant and then happily trots off into the arms of a cougar, it is unlikely to return to pester the plant.")

Other toxins hold nutrients hostage, gridlocking digestion. Tannin, for instance, binds peptides (the building blocks of proteins) so tightly that they can't be teased out by the digestive enzymes that normally disassemble food. Other toxins work by pinning the arms of these digestive enzymes. Either way, the protein remains unbroken and unused, and you go hungry. The only way to loosen the grip of digestion inhibitors is to heat the offending plant to 100 degrees Celsius, which is why the discovery of fire gave early humans a truly

Promethean power. As a wild herbivore, however, you can't crank up the Jenn-Air range when faced with a suspect plant. You have to detoxify the poisons in plants internally, using your own chemical laboratory. In the end, nutrition becomes a wrestling match between the plant's chemical profile and your physiology.

Things get really interesting when the plant changes its chemical profile. When stressed by poor soil or moisture loss, for instance, a plant may beef up its chemical arsenal, not wanting to lose even a single leaf. Depending on the terrain, one tree might be fine to eat, while the same species in a patch of poorer soil may hold a bitter harvest. The very act of puncturing a leaf may cause a tree to fight back with an overproduction of toxins—changing its chemistry in as little as forty minutes to protect the rest of its leaves. As a herbivore, you never know what you're going to get, not from forest to forest, tree to tree, or even from one side of a tree to the other.

Even if your system is equipped to detoxify a toxin, it takes energy to boot offending molecules out of your liver, repair your DNA, launch antioxidant artillery, or shed the poisoned cells in your mouth, stomach, esophagus, intestine, and so on. If you spend more energy breaking down or flushing out a toxin than you receive from the food itself, you could rack up a negative nutrient score. But if you don't detoxify secondary compounds, they'll filibuster your digestive system. Either way, you may look like you're feasting, but you could be steadily starving. And natural selection, which deals harshly with foolish genes and foolish choices, won't let that go on for long. An out-of-balance diet will eventually weaken you, and those genes (your genes!) will be edited out of the population.

If you want to keep your place in the gene pool, explains Glander, there are at least three dietary strategies you can adopt. You can be a specialist like the eucalyptus-chomping koala—eating one plant that your whole digestive system is dedicated to detoxifying. Or you can be a generalist and eat small amounts of lots of different species, so your body has to detoxify only tiny batches of toxin, spreading out the risk. Or you can do what our primate ancestors did: eat a limited selection of plants but be very picky—selecting only the choicest parts of the plants so that you net more nutrients than toxins.

Once upon a time, before there were nutritionists or USDA safety inspectors, our primate ancestors knew how to put together a sensible, safe diet. Somehow, they'd learned to shop the supermarkets of the plains, jungles, and seas, avoiding the dangers while cash-ing in on digestible nuggets of nutrition. In a country where millions are spent each year on diet and nutrition advice, why haven't we consulted the mammals, birds, and insects that successfully act as their own nutritionists? Might their choices show us what we may have been *meant* to eat, in a purely biological sense?

SMART EATING: WILD CONNOISSEURS

Strangely enough, not much research has focused on the chemical intricacies of animal food choice. Glander, one of the few primatologists who has published on the subject, devised a way to demonstrate the nutritional common sense of the new guests (*Lemur fulus*) at his Primate Center. "Before placing them out in the forest, I gave the lemurs ten leaves—leaves from local species like sweet gum that they had never seen before. I made sure there were no outright poisons (not wanting to take any chances with an endangered species), but I did include five leaves that contained digestion inhibitors and five without. After sniffing and puncturing like trained tasters, they spit out the bad and swallowed the good. Their menu was a balanced mix of leaves with the highest digestibility, the highest nutrient content, and the lowest tannin content. We couldn't have hired a nutritionist to do a better job."

Glander's hunch about primate palates was developed while studying the discriminating tastes of mantled howler monkeys (*Alouatta palliata*)—a tree-dwelling species native to Costa Rica, Panama, and Mexico. He had followed the howlers day after day, watching them move through the jungle like picky eaters at a dinner buffet—eating only certain leaves, or certain parts of leaves from one tree, while ignoring a neighboring tree of same species. To find out why, he chemically probed both the plants they ate and those they passed over. Turns out the plant material they avoided was either full of alkaloids and condensed tannins (especially virulent protein hoarders) or conspicuously low in protein and unbalanced in amino acid counts. He concluded in his paper: "Howlers are chemically astute; they consistently chose material of the highest nutritive value and passed up the low-value material with secondary compounds."

Katherine Milton, a professor of anthropology at the University of California, Berkeley, would have to agree. She studied howlers on Barro Colorado Island in Panama, examining what age leaves they preferred. In her 1979 study, she found that howlers overwhelmingly preferred young leaves to older ones, perhaps because the young leaves earn them a higher return of energy per unit of weight.

In a 1978 study, Doyle McKey and his colleagues reported a similar cautiousness among black colobus monkeys (*Colobus satanus*) in the Douala-Eden Reserve in Cameroon. The monkeys in this particular reserve studiously avoided a common tree species that colobus happily ate in other parts of the country. McKey guessed that poor soils at the reserve must have made the native tree species beef up with toxins to protect every hard-won leaf. Sure enough, when he analyzed the avoided leaves in that region, he found them loaded with phenolics, which are digestion inhibitors. The only part of the plant the colobus would eat was the seeds, which were high enough in protein to be worth the trouble of detoxifying.

On the trail of yet another instance of smart eating, Harvard anthropologist Richard Wrangham and his colleague Peter Waterman looked at vervets feeding on acacia trees. The vervets lustily devoured immature leaves, seeds, fruits, and flowers of two species of acacia (*Acacia tortilis* and *Acacia xanthophloea*), but when it came to eating the gum, they got picky. Only gum from *A. xanthophloea* was eaten, while the reddish-brown gum of *A. tortilis* was completely ignored. Analysis showed that the ignored gum had high levels of condensed tannins and not enough protein to justify the effort. The sought-after gum, on the other hand, was high in soluble carbohydrates and devoid of tannins. A hassle-free wad of energy.

Wild baboons (*Papio anubis*) of Africa also seem to know how to stretch a feeding dollar. Researchers Andrew Whiten and Dick Bryne of Scotland's University of St. Andrews found that the baboons preferred plants or plant parts that were high in proteins but low in hard-to-digest fiber and alkaloid toxins. When the baboons had no choice but to eat a high-toxin menu, they made sure they picked individual plants that were also rich in proteins. Plants that were lower in proteins—not worth the trouble—were simply passed over.

As you can see, smart eating is more than just avoiding or minimizing encounters with nasty toxins; it's finding the proper mix of nutrients and building blocks that the body needs. It seems that animals have a nose for what is good for them and actually crave it.

Cravings

In his famous "cafeteria" studies of a half-century ago, Johns Hopkins's Curt P. Richter broke rat chow into its constituent parts and placed them in eleven separate dishes: proteins, oils, fats, sugars, salt, yeast, and so forth. Given unlimited quantities, rats mixed and matched, procuring a diet that, with fewer calories, allowed them to grow faster than rats fed the normal chow. In fact, the nutritionists were surprised by the choices—they finally had to admit that the rats had composed a better nutritional diet than the makers of the rat chow!

Scientists think that a craving for a complete diet may have also influenced how America's great buffalo herds moved across the landscape. One theory holds that their traditional routes purposely included salt licks and other reliable sources of vital minerals. The continual roaming also may have helped the bison avoid grass tetany, a springtime malady that affects fenced livestock. Turned out to fresh green fields, horses and cows sometimes binge on immature grass, which is high in nitrogen and potassium, but low in available magnesium. If there are no sources of magnesium in the field, the springtime "feast" can leave livestock with the "grass staggers" or even kill them. If livestock have the opportunity to range, however, they will avoid grass tetany by balancing their nutrients. In the same way, white-tailed deer will assiduously search for a balanced diet, moving methodically through the woods and fields, cobbling together the nutrients they need. Bucks are even choosier than does, searching for plants that contain enough potassium, calcium, and magnesium to fuel their fantastic spurt of antler growth. "This nutrient-specific eating looks pretty smart to us," says Bernadette Marriott, behavioral ecologist and deputy director of the Food and Nutrition Board at the National Academy of Sciences. "We could use some lessons."

I found Marriott in a chic office building tucked near Canal Park in Georgetown, a plum location for the National Academy of Sciences. The security was oddly strict, but Marriott was welcoming. Petite, dark-haired, and dignified, she impressed me as someone who might have been shy at a former stage of life, but who has now turned her shyness into a personal power that says what is important, but will not shout to be heard. From a window that spanned the width of her office, a corona of afternoon light framed her head and streaked into her soothingly darkened office. Behind her on a credenza were photos of Himalayan peaks and batiks of many-armed dancers. On her desk, a figurine of India's rhesus monkey—the species she had studied—held her business cards up to visitors.

Like many biomimics I talked to, Marriott was drawn to the zone between two disciplines. To satisfy an interest in biology and psychology, she decided to study why animals choose the foods they do and what effect this has on social evolution. "I studied rhesus monkeys [*Macaca mulatta*], which are phenomenally finicky eaters. They spend a lot of time preparing food to eat, stripping off the edges of a leaf, or eating just the midrib. I wondered: How do they learn and remember which foods are safe and nutritious? Are they keying in on color, shape, texture, or is it something else?

"When I watched and analyzed behavioral patterns, it turned out that shape—which I thought might be the key to the search image—was not statistically important. This led me into the chemistry lab to do a nutritional analysis of everything they ate. [Like Glander, she was delving into unexplored terrain here.] What I found astounded me. These monkeys managed to pick a diet that was perfectly balanced. The only thing lacking, however, was some minerals that they needed."

Looking back on it, knowing rhesus needed minerals, she says she shouldn't have been surprised to see them eating soil. "As a Westerner, your first instinct is to say, don't put that in your mouth. But we knew from their behavior that it wasn't a mistake—it was something important." The monkeys make a special trip to a particular cliff, where they scratch the soil with their fingers and then eat it. After many years of use, an actual cave is formed, big enough for a monkey to stand in. Usually, one site is used religiously by the entire troop. When researchers picked sites at random and tried starting caves for the monkeys, the rhesus came and investigated, but wound up back at their own site. They would actually form lines outside and wait their turn rather than go somewhere else and start digging.

Once she began asking, Marriott learned that many people in Africa eat dirt, as do people in this country. "It's called geophagy, and in the United States, it's very covert, taboo. Whenever I speak on this subject, people come up to me and tell me they have an aunt or a neighbor who eats dirt. It's never *them*, of course," she says with a slight wink.

Turns out there's an industry in dirt eating. In the Italian markets of Philadelphia, you can buy commercially produced cakes of soil stamped with insignias of their origin. "Georgia dirt is supposed to be prime," says Marriott, "but when I have tried asking merchants what it is, they simply say, 'It's good for you. It'll help you have strong babies.' They never say it's dirt."

When Marriott first watched the rhesus eating soil, she thought maybe they were after bugs or tubers, but an analysis of the soil showed nothing of the kind. Donald E. Vermeer of George Washington University in Washington, D.C., theorized that the dirt might chemically bind to and neutralize stomach acids, helping to quell stomach upset. Sure enough, his structural analysis showed the presence of kaolin, which is the active ingredient in Kaopectate. Timothy Johns, a biochemical botanist at the University of Toronto and author of *With Bitter Herbs They Shall Eat It*, believes that the benefits of dirt are more physical than chemical. He thinks clay particles in the soil physically bind to the secondary compounds in ingested plants, thus occupying them so they cannot be incorporated into the body. Johns bases his belief on the observation that Bolivian Indians coat their wild potatoes (which are full of toxins) with a slurry of soil before cooking.

Marriott departs from both the chemical binding hypothesis and the physical binding hypothesis. "I came away with the theory that geophagy is more a quest for something good than a way to rid the body of something bad. I think the rhesus eat soil as a broad-based mineral supplement. While clay or kaolin may provide a feeling of well-being—by settling the stomach—that's perhaps a secondary benefit acting to reinforce the mineral-gathering behavior."

To verify what she suspected, Marriott took soil samples back to her lab and analyzed them. Sure enough, soils from the troop's traditional feeding places showed spikes in certain minerals like iron that the monkeys were missing in their diets. Might the monkeys be waiting in line for their one-a-day mineral pills? Marriott smiles and shrugs. "At least they don't have to pay sixteen dollars a bottle."

Are You What You Eat?

It's easy to understand how safe eating would evolve, but how about smart eating? Are those animals that can locate a particularly rich form of fat, protein, or mineral being rewarded in some evolutionary way? Michael Crawford and David Marsh, authors of *The Driving Force: Food, Evolution, and the Future*, argue that evolution is indeed substrate-driven and that the key substrate is food. If you want a new and improved body to put you in a better survival position, they say, you have to first snag yourself the building blocks to make that change.

Morphologists tell us that certain body structures would be impossible to build without enough of the right kinds of food. To build a brain, for instance, you need miles of lipid (fatty) membrane to wrap around neurons, and lots of vascular tissue to feed those neurons. Both components are made of essential (long-chain) fatty acid derivatives, which are chemically manufactured in a herbivore's body, starting with the fats in leaves and seeds. An easier way to amass large quantities of these "neural" fatty acids is to eat animals that have already manufactured them for you. Switching from a leaf-only diet to one with meat, therefore, might have given carnivores a larger supply of neural building blocks—the ticket to advanced structures like keen eyes and a bigger brain.

Second, in addition to being a structural material, food is also a batch of chemicals, which, by their nature, are reactive. When these substances enter the body, they bump into and interact with the bath of hormones, enzymes, genes, and neurotransmitters that govern and regulate cell life. Above a certain threshold concentration, food chemicals may begin to influence which enzymes start to work, or when genes will turn on or turn off.

This threshold mechanism gives food the ability to tweak powerful control knobs within the body. Imagine, for instance, that an adaptation is lying dormant in the genes, just waiting for a chemical surge to "turn it on." There's no telling what might emerge as a result of a good diet. Witness the spurt in human height, for instance, when nutritious foods became widely available in the Western world. In this case the nutrients affect the phenotype (the growing body) but not the genotype (the set of instructions encoded in DNA that is passed from generation to generation). Take the diet away from the next generation of phenotypes, and heights will shrink to prior averages.

But what if diet can affect certain aspects of our permanent genotype over the long haul? Crawford and Marsh think that it can, and they offer the following rationale. If you can eat an animal that makes an important nutrient, such as vitamin A, you no longer have to devote your biosynthetic pathways to making vitamin A. This frees your energy for other chores, like building a brain. It may also free up genetic space, the authors speculate. Say you have only so much room on your chromosomal "hard drive," and it's already filled with genetic instructions. By eating vitamin A manufactured by another animal, your instructions for synthesizing vitamin A become superfluous. If a mutation suddenly rewrote that gene sequence with another set of instructions—a new adaptation—you wouldn't miss the vitamin A recipe, and you could therefore live to take advantage of, and pass on, the new adaptation. Evolution, stuck on its plateau, would suddenly spring to a new level.

If this theory is even a little bit true, you can see how important it is for an animal (and for us) to have the good sense to gather what is needed in terms of food. But where is the command center for our fine gastronomic compasses? Is our good taste hardwired into our bodies or is it learned? The researchers I talked to think it might be a little of both.

How Did Smart Eating Develop?

The first primates were exclusively insect eaters, Glander tells me. By eating insects that fed on plants, the primates were ingesting plant compounds by proxy. By the time the primates evolved into plant eaters themselves, they had already developed the physiological apparatus to either metabolize certain nasty plant chemicals or to excrete them. Because plant poisons vary from plant to plant, however, these "safe plants" would be a small subset of the whole. If a primate wanted to step out of this limited range and try others, it would need some way to determine what was good and what was vile. Luckily, a knack for smart eating develops in two ways. It's partly hardwired into our senses by evolution, and partly acquired or learned through life.

Glander is one of many researchers who suspect that the main form of primate leaf discrimination is through the senses of taste and smell. When the lemurs tasted the trial leaves, they sniffed and sometimes took a leaf into their mouths and punctured it, allowing the volatile compounds to waft over their Jacobson's organs—the interconnected passageway between the mouth and the nasal passages. Presumably, it is in these smell/taste receptors that chemical analysis occurs.

As mammals, we can sense bitter, acrid, astringent, sour, and pungent flavors—all of which serve a function in food selection, says Richard Wrangham of Harvard University. Consider sourness, for instance. Sourness is a measure of acidity, which acts as a natural preservative against bad microbes (the ones that cause rancidness). Somewhere deep within us, we recognize sourness as a badge of purity, assuring us of a food's safety. That may be why we prefer a little sour flavoring in our sweet confections, rather than straight sugar.

Certain kinds of fermentation—when a fruit turns to alcohol, for instance—may also signal safety to an animal. Fermentation in fruit is assisted by bacteria that deactivate unpleasant compounds such as cyanide and strychnine. On the other hand, there is also bad fermentation—the action of different kinds of microbes whose metabolic waste is toxic, even deadly, to humans. To avoid them, we are hardwired with a strong aversion to rancid flavors.

Our hardwiring is not absolute, however, and our aversion to or craving for certain foods may sometimes become curiously strong. In her book *Protecting Your Baby-to-Be*, evolutionary psychologist Margie Profet suggests that pregnant women's unusual taste swings may be adaptations designed to protect embryos during sensitive development cycles. If true, this could explain everything from morning sickness to the pregnant woman's inexplicable love affair with pickles. Perhaps the real appeal of pickles is their sourness, says Profet, a badge of purity at a time when rancidness must be avoided. Later in the pregnancy, the woman may begin to crave what she's missing nutritionally—a specific hunger being hardwired into her neurons right on the spot. Thomas Scott of the University of Delaware found that when a rat is deprived of salt, neurons that normally respond to the taste of sugar are commandeered and reprogrammed to become receptive to salt. In other words, salt becomes as pleasurable to the brain as sugar normally is. Cravings might also be heightened through our other senses. When we're hungry, for instance, the brain throws open the olfactory receptors, making us more sensitive to the odors of food. (That's why head colds and smoking suppress appetite—we can't smell our food as well.)

Even this flexible hardwiring can't fully explain the fine discrimination shown by animals, however. Regardless of how plant-smart their inborn sensors are, nothing could prepare an animal to automatically recognize every species in the jungle. Some things just have to be learned on the job.

With primates (and many other animals, such as elephants), the learning begins with Mom. Infants will peer and poke into their mother's mouth to smell and taste what she is eating, and after a while, they build a chemical profile of what's good. "It's like downloading information from a computer," says Glander.

Once they leave their mother, primates have to keep on making decisions about whether new foods they encounter are safe and worth collecting. Using themselves as guinea pigs is one option, but social primates have found a better way. Kenneth Glander calls it "sampling." When howler monkeys move into a new habitat, one member of the troop will go to a tree, eat a few leaves, then wait a day. If the plant harbors a particularly strong toxin, the sampler's system will try to break it down, usually making the monkey sick in the process. "I've seen this happen," says Glander. "The other members of the troop are watching with great interest—if the animal gets sick, no other animal will go into that tree. There's a cue being given—a social cue." By the same token, if the sampler feels fine, it will reenter the tree in a few days, eat a little more, then wait again, building up to a large dose slowly. Finally, if the monkey remains healthy, the other members figure this is OK, and they adopt the new food.

Not all monkeys volunteer for sampling duty, however. Glander has noticed that monkeys in vulnerable stages of their lives—juveniles, subadults, and lactating or pregnant females—seem to bow out of sampling. If the risks are too great for some monkeys, why would any monkey volunteer? "I think the benefits may be genetic," says Glander. Adult monkey fathers, for example, may be boosting the health of their offspring by testing foods for their pregnant or lactating mates. Adults that aren't yet parents may also volunteer, pointing out wholesome foods for their siblings and nieces and nephews who share a portion of their genes. Despite these benefits, Glander says no monkey would want to risk being a full-time sampler. "The sampler role shifts from monkey to monkey, so as to spread the risk and not unduly jeopardize anyone. This risk-sharing is, in itself, a good reason for being social," speculates Glander. Sampling, he believes, may have in fact contributed to the development of social behavior in primates.

Besides tipping the scales toward sociability, tricky food choices may also have challenged animals in ways that rewarded intelligence. Researchers hypothesize that sometime in the Middle Miocene (7 to 26 million years ago), monkeys developed the ability to tolerate higher levels of toxins than apes could, giving monkeys a wider choice of foods. Apes (our ancestors) were stuck with a more sensitive digestive system and were therefore forced to roam in search of higher-quality foods and new ways to prepare that food. Richard Wrangham believes this may have contributed to our ape ancestors finally leaving the jungle, walking upright onto the plains, and beginning to use tools and fire.

The droughty climate of the apes' new plains habitat meant that foods were more seasonal in nature. To find reliable nutrition throughout the year, they had to problem-solve, employ tools, and perhaps cooperate more with their fellow primates. As it turns out, although monkeys won the evolutionary race to detoxify compounds, apes wound up with higher mental functions.

Female apes were faced with even more limitations and nutritional demands. Unlike males, who could squeak by on lower-quality foods or take excursions to far-flung corners of their habitat for pockets of early ripening fruit, females were often eating for two or lactating. They needed safe, nutrient-rich, protein-rich, calcium-rich foods, but they couldn't travel far to find them. Faced with this dilemma, females may have been the first to experiment with new types of foods, such as flowers, young leaves, and tubers, and to experiment with hand-held tools. Michelle L. Sauther, an anthropologist at Washington University in St. Louis who has studied food choice in primates, writes, "[Ape] females may have broken free from some of the seasonal constraints on food availability by using tools to gather wild plants, insects, and small mammals. For example, females may have employed digging sticks for underground tubers and used techniques similar to those observed in wild chimpanzees, such as using stone hammers to crack open nuts and employing termite and ant wands [sticks thrust into hives and nests to harvest insects]."

Did resourcefulness like this advance the whole era of tool using? Sauther concludes that the responsibility of being a mother may not have been a burden but, rather, a "catalyst for developing more efficient foraging techniques." A sensitive stomach, a new habitat, and the hungers of pregnancy were perhaps the literal mothers of invention. Chances are also good that those females who were very good at finding year-round food in seasonal habitats went beyond "just surviving" and began to tap the power of limits. Branching out from their standard fare, they may have actually garnered better nutrients and therefore provided their young with the metabolic stuff needed to develop a bigger brain.

Many millennia later, what have we done with all those smarts?

What Can Animals Teach Us About Smarter Diets?

It appears, as the whole country idles in the drive-through lane of Burger World, that we've lost our dietary way. Even the most nutrition-savvy of us can have a hard time keeping Oreos in the house for more than a few days. In America, where 30 percent of the population is obese and suffering from diseases that are aggravated if not caused by poor diets, we could use a crash course in choosing nutritious foods.

What's strange is that feeding behavior—specifically food selection—is one of the last things that has been examined in both human and nonhuman primate studies. For all we know (and we don't), humans may have originally learned to gather food by watching what other primates ate. Today, there are still some overlaps in the diets of human societies and animals in the same habitat. Says Bernadette Marriott, "Many of the foods that monkeys in Nepal gather are ones that people also gather, although those practices are going by the wayside as we introduce people to commercial foods. After doing nutritional profiles on these [native] foods and seeing how rich they are, we are now trying to encourage people not to give up on their wisdom, to eat more of these widely available plants rather than buying Western food."

In many places, we are too late; people have already gotten away from eating what the animals eat. The Green Revolution of the 1960s "converted" whole nations from a relatively healthful, native-derived crop diet to one of foreign-bred wheat, rice, corn, oats, and so on. Everywhere, farmers have abandoned local plants that were hardy, disease resistant, and well-suited to their climate, and are instead growing plants imported from other regions, plants dependent on chemical and petroleum companies for their yields.

Now the cycle is coming back around—after dangerously homogenizing our crops, we are reevaluating wild varieties. It might make more sense, we're admitting, to grow yams from local breeding stock than to import Idaho potatoes that have half the flavor and twice the water and pesticide requirements.

To round out the crop choices for more bioregional agriculture, we may want to enlist the help of animals that have already forged a clear path through the chemical jungle. Unhampered by the blinders of human custom, they might lead us to crop lines that, though new to us today, are old standbys of the primate clan.

MEDICINAL EATING: ANIMAL PHARMACISTS

From here, it's not such a long leap to imagine that animals might have *more than one relationship* with the green plants that surround them. They might, for instance, regard certain plants not as food but as medicine that helps them feel better when their systems are out of whack.

"Out of whack" can mean that parasites have come to call, or that bacteria are multiplying. Plants fend off bacteria, viruses, roundworms, nematodes, or fungi by producing secondary compounds. If these compounds were placed inside an animal gut, might they offer the same antibacterial, antiparasitic, or antifungal protection? And might the animals notice this and seek out treatments when needed? After all, if they've learned to avoid toxic secondary compounds, couldn't they just as easily learn to harness the beneficial ones?

Animals acting as their own pharmacists. We shouldn't be surprised, but we are.

Plop, Plop, Fizz, Fizz

From where Michael Huffman sat, relaxed, primatelike, slanting glances from his peripheral vision, he could see the chimp they called CH. As we humans would say, she was not herself. Lethargic, unable to rouse herself from her tree nest, CH hardly noticed the intense feeding activity all around her. Over the last few days, her urine had been dark and her stools infrequent and irregular—classic symptoms of roundworm or schistosomiasis infection. That morning, CH suddenly struggled up and staggered off into the jungle of Mahale Mountains National Park, located on the eastern shore of Lake Tanganyika in western Tanzania. Huffman, a primatologist from Kyoto University in Japan, and Mohamedi Seifu, of the Mahale Mountains Wildlife Research Center, grabbed their notebooks and followed.

Chimpanzees (*Pan troglodytes*) have an uncanny knack for finding their way in the jungle and for remembering and anticipating ripening food trees. This orienteering ability, as behavioral biologist Richard Estes calls it, allows them to get from wherever they are to wherever they want to be via the shortest route. Even though she was ill, CH seemed to know exactly where she was going, and she didn't stop moving until she reached the flowering *Vernonia amygdalina* shrub. It was a plant that chimps don't normally eat, but one that people in many parts of Africa use for traditional medicine. With painstaking care, she selected several young shoots and began to strip off the leaves. Using her front incisors, she peeled back the bark, exposing the liquidy pith. Grimacing like a coed downing a shot of tequila for the first time, she chewed the branches and sucked out the juice.

Huffman watched CH carefully after her "treatment," and, sure enough, within twenty-four hours she was defecating regularly, foraging for longer periods, and eating with the rest of the troop. When chemists later tested the plant, they found two secondary compounds in the pith—sesquiterpene lactones (terpenes) and steroid glucosides—both of which were shown to have exceptional antiparasitic activity, strong enough to kill a wide variety of gut parasites without killing the patient. Sometime later, Huffman was lucky enough to see a second chimp seek out the pith. This time, he was able to monitor the chimp's parasite levels (by checking feces), watching them drop to harmless levels within twenty hours of treatment.

Once he knew what to look for, Huffman realized that a lot of chimps were using *Vernonia* pith, especially during the wet season when worms are abundant. Despite the fact that this particular species of *Vernonia* plant is rather rare in the Mahale Mountains, both the chimps and the native people have zeroed in on it. *Vernonia amygdalina* is called "bitter leaf" by the Tongwe natives, who use it when they are afflicted with similar malaise, loss of appetite, and constipation. The pith contains a perfect dose of the juice, about the same amount as in a typical dose used by humans. Further analysis revealed why the chimps focus only on the pith—elsewhere in the plant, in the leaves and bark, for instance, the parasite-slaying toxins are in concentrations high enough to kill lab mice.

Encouraged by the antiparasitic qualities of this one plant, researchers have begun to investigate the entire *Vernonia* genus. Clinical tests of a closely related plant (*V. anthelmintica*) have yielded a compound that could be used to treat pinworm, hookworm, and *Giardia lamblia* in humans. "Standard wisdom is that these [secondary compounds] are toxic or dangerous to animals," writes Richard Wrangham. "But over the last fifteen or twenty years, a series of anecdotes has jelled into studies suggesting that animals can use those compounds to their own benefit, often turning the toxic effects against their own internal enemies." So much for standard wisdom.

Take Two Leaves

Another clue to the puzzle showed up a few miles from Huffman's post, in the Gombe Stream National Park in Tanzania. A troop of chimps living in Gombe are among the most scrutinized animals in animal behavior history. For more than three decades, primatologist Jane Goodall has trained many an observer there, including Harvard anthropologist Richard Wrangham. Wrangham says he became a believer in animal self-medication when he witnessed something one early dawn, before any of the other researchers were on their "beat."

"A chimp I was observing had woken up sick," he tells me, "and instead of rolling over for more sleep, she got up and began walking, making a beeline really. I had to hustle to keep up with her. Twenty minutes later she stopped at an *Aspilia* plant [a cousin of the sunflower that grows as high as six feet] and began a most unusual ritual." The chimp began carefully inspecting certain leaves, even holding them in her mouth while they were still attached to the shrub, abandoning those that didn't suit her. Finally, she plucked a small leaf and tucked it under her tongue, the way we might pop a nitroglycerin pill. She let it linger there, rolling it back and forth a little, but not chewing. Richard wondered if she might be absorbing something from the leaf through the mucous membranes under her tongue.

From his hiding place, he watched in amazement as she puckered up her face and swallowed the hairy leaf, which must have been like swallowing a patch of fuzzy leather. He watched her swallow a dozen more leaves at a slow rate (five leaves per minute swallowed as opposed to the normal thirty-seven leaves a minute for leaves that are chewed) before she moved back to the troop.

It was obvious from her grimace that this was not a taste treat, but Wrangham couldn't automatically assume it was medicinal either. "Feeding studies are tricky," says Wrangham. "It's not enough to check 'eating' or 'not eating.' You have to catalog which chimp is eating which leaf from which plant, and then count exactly how many leaves it eats." Even then, as Karen Strier, an anthropologist at the University of Wisconsin in Madison, reminds me, you may not have any useful information. "The digestive tract is a black box," she says. "You don't really know what the animal 'makes' of what it eats—whether compounds are absorbed or destroyed in their journey through the body. Your only clue is to analyze what's left of the food—what comes out the other end in the feces." Indeed, what remained in the feces after the strange leaf swallowing—a handful of nearly intact green leaves—become a signature clue for Wrangham. If the leaves were not digested, then what purpose were they serving?

Though chemical analysis of the ingested leaves showed no conclusive proof of "medicine," Wrangham began to see more and more of the strange leaf-swallowing behavior. One troop of chimpanzees in Kanyawara, a community of Kibale Forest National Park in western Uganda, seemed to be increasing their intake of leaves during certain times of year. Sure enough, when he looked over a sequence of months, he saw that the spike in leaf-swallowing behavior coincided with the months of heaviest tapeworm infection. This was the first time that leaf swallowing was correlated positively with a specific parasite infestation. Wrangham also noticed that the dungs with whole leaves in them also contained tapeworm fragments. It seemed as if the leaves, hairy and whole, might have caused a motile fragment of tapeworm to be shed from the gut and then carried off with the feces.

Meanwhile, in the Mahale Mountains, Huffman was also finding spikes in leaf-swallowing behavior during the rainy season, when loads of parasitic nematode worms tended to be higher. Could it be that the chimps were downing more leaves at that time for the same reasons we buy more cold medicine during the cold and flu season?

The latest theory is that the abdominal pain caused by nematodes or tapeworms causes chimpanzees to increase leaf swallowing, just as a tummy ache might cause your dog or cat to go out and eat grass. What researchers don't yet know is whether the worm-purging effect is chemical (worms repelled by medicinal compounds) or mechanical (worms being combed out of the gut by the hairy leaves). Nevertheless, something about *Aspilia* seems to be affecting parasites, and the chimps know that.

To find out what else they know, researchers are now looking for other plants swallowed whole by primates. In a chapter of the 1989 book *Understanding Chimpanzees*, Richard Wrangham and co-author Jane Goodall report that Ugandan chimpanzees have been seen swallowing the leaves of the *Rubia cordifolia*. Of the 401 chimpanzee fecal samples he collected in Kibale, Wrangham found *Rubia* leaves in 16. All were whole and without tooth marks—signs of the same down-the-hatch fate that befalls *Aspilia* leaves. An analysis of the leaves uncovered a triterpene called rubiatriol, some bioactive anthraquinones, and most exciting of all, a cyclic hexapeptide that is "an extremely potent cytotoxic agent which is being investigated by the National Institutes of Health as a therapeutic agent for cancer patients."

Suddenly, with the verified connection to possible cancer-fighting ability, these compounds found in a far-off jungle were no longer molecular footnotes. And the grimacing feeding sessions were no longer anomalies. It was time to put the self-medication anecdotes to the laboratory test.

First on the list was *Ficus exasperata*, which is thought to kill nematodes, an important intestinal parasite of chimps. The chimps concentrate on the young leaves, which have six times as much of the active compound (5-methoxypsoralen) as the old leaves. According to Eloy Rodriguez, a plant biochemist at Cornell University, the leaves and fruits of *Ficus* do a good job of killing the food-poisoning bacteria *Bacillus cereus* without harming *Escherichia coli*, the good bacteria that live in the gut. Many more leaves are waiting for chemical examination. Among the fifteen plants shown to be swallowed, not chewed, are *Aneilema aequinoctiale*, *Lippia plicata*, and *Hibiscus aponeurus*. Researchers are also collecting any plants that are eaten only on rare occasions or that are rubbed on the animal's fur instead of being swallowed.

Wrangham's next big project is a study of diet differences between monkeys and apes such as chimps. As mentioned earlier in this chapter, monkeys can tolerate secondary compounds better than chimps can. Therefore, says Wrangham, "watching what monkeys eat and what chimps avoid may lead us to some interesting secondary compounds—possible drugs." Plants that both species avoid are likely to be loaded with secondary compounds, substances that even local healers may not know about. The only problem with this approach, Wrangham tells me, is that it may have come too late for many species of plants. "Every time you take a leaf in to be analyzed," he says, "you wonder if you'll be able to find the species in the wild again."

Awash in Evidence

Why have we waited until it is almost too late to start this quest? The early 1980s was the first time scientists speculated (in print at least) that primate leaf-swallowing behavior might be connected to self-medication. And yet we've known for a long time that rats "treat" themselves by swallowing clay after ingesting poisonous amounts of lithium chloride. In fact, experiments have shown that if the rat even *thinks* it was poisoned, it will eat clay, which is thought to absorb the toxic load. In the same way, as every pet owner knows, when a dog takes itself outside for an aperitif of grass, it is looking to purge what ails it.

"Why we thought that hominids were the only ones who could discover the curative properties of plants, I don't know," says Wrangham. "We're not the only animals in the jungle." Wrangham also figured he was not the only researcher who had noticed animals self-medicating. When he and Eloy Rodriguez decided to hold a symposium at the 1992 American Association for the Advancement of Science (AAAS) meeting, scientists came out of the woodwork with their stories. The field of zoopharmacognosy was born.

At that meeting, Jane Phillips-Conroy of Washington University in St. Louis gave an account of baboons near Awash Falls in Ethiopia that live in the ideal "controlled" experiment, set up by geographical differences in their home ranges. Two populations of the same species of baboon (*Papio hamadryas*) live near Awash Falls; one population feeds exclusively above the falls, the other below. The population below the falls is vulnerable to a snail-borne schistosome (*Schistosoma cercariae*), a fluke worm that causes a debilitating disease in primates, including humans. Above the falls, the snails are free of the fluke worms.

Also distributed above and below the falls is *Balanites aegyptiaca*—a plant whose berries and leaves contain a steroidal saponin called diosgenin, a compound known to be active against the fluke worm. Native peoples have long used *Balanites* for controlling infections of schistosomiasis, and so, it would seem, do baboons. In fact, although both populations of baboons have access to the healing plant, the only baboons that eat it are the ones that live with the infected snails. This led Phillips-Conroy to speculate that the plant was being sought out for something other than nutritional purposes, or else both populations would partake of it.

Another tale was told at the meeting about two populations of mantled howlers, tree-dwelling monkeys that are habitually plagued by parasites. Researchers in Costa Rica were surprised by the stark contrast in parasite loads between two populations living in different parts of the tiny country. Howlers in Hacienda La Pacífica were heavily parasitized, while howlers from Santa Rosa National Park carried surprisingly light loads. Searching for reasons, the researchers noticed that Santa Rosa, with light parasite loads, had plenty of fig trees (*Ficus* spp.), while La Pacífica had none. Knowing that humans use the latex in fig trees as an antiworm medicine, researchers at the conference theorized that a compound in fig leaves or fruits may be keeping worm loads under control in the Santa Rosa howlers.

Another unusual finding was the howlers' utter lack of gum disease or tooth decay. Could howlers be brushing and flossing regularly? More likely, say researchers, it has something to do with the pedicels (stalks) from the cashew (*Anacardium occidentale*) that they are known to eat. An analysis of the pedicels showed high amounts of the phenolic compounds anacardic acid and cardol, both of which kill gram-positive bacteria such as *Streptococcus mutans*—the critters that cause tooth decay in humans.

Also discussed at the meeting was the whole realm of medicines applied in non-oral fashion. Anecdotes abound about birds such as eagles that line their nests with resin-soaked pine sprigs, perhaps to keep out nest parasites. The blue jays on your front lawn may also be practicing a form of medicine. In a ceremony called anting, jays squeeze ants in their beaks and then rub the formic acid onto their feathers. They seem to have an almost beatific look on their faces as they do this, as if the ant juice is intoxicating. Other investigators have postulated that anting is actually an antiparasitic gesture—a delousing.

Bears are also known to exhibit strange rubbing behavior. After spending seven years with a Navajo family and learning about traditional tribal medicines, Harvard ethnobotanist Shawn Sigstedt became intrigued by the fact that there were so many medicinal plants with names that included "bear." Traditional Navajo teachings said that medicines were given to people by the bears, a good indicator that the Navajo might have watched animals self-medicating and then adopted their practices. Sigstedt put the bear connection to the test with *Ligusticum porteri*, a vanilla-celery-scented herb that grows in the Rocky Mountain and Southwest regions of the United States and is used by the Navajo to treat worms, stomachaches, and bacterial infections. He gave samples of the plant to polar bears and grizzly bears in the Colorado Springs Zoo, and watched in amazement as they rolled and rubbed with relish, perhaps getting relief from ticks or skin fungi.

How Did Animals Learn to Self-medicate?

In a way it seems contradictory. How could the practice of eating toxins to self-medicate have evolved, when there is so much evolutionary pressure *not* to eat toxins? As with safe eating, says Richard Wrangham, there are probably physiological, behavioral, and cultural aspects to the phenomenon of curative eating.

First the physiological. Depending on what an animal needs in its diet, even hardwired tastes can reverse themselves. When an animal wakes up sick, for instance, it might find its aversion to secondary compounds transformed into a tolerance or even a craving for bitter leaves. Chinese herbalists have used this body feedback for thousands of years when treating human patients. As Michael Huffman reported at the AAAS conference, "Sick people supposedly tolerate a level of bitterness utterly repellent to healthy individuals, allowing the herbalists to identify when and how to adjust their doses." The sicker a patient is, the more bitterness he'll tolerate. When he starts complaining that the medicine is too bitter, the herbalist pronounces him cured.

Jane Goodall has experimental evidence that seems to support this theory. When Goodall needed to treat some chimpanzees with tetracycline, a bitter substance, she hid it in bananas, and watched who ate what. While the healthy chimps turned up their noses at the laced treats, the sick chimps quickly ate the bananas, seemingly oblivious to the bitterness. In wild feeding, Huffman found that chimpanzees with the highest parasite loads tend to eat the bitterest leaves. Glander observed a similar flip-flop in howlers' tastes. Healthy howlers avoid leaves that are high in tannins and therefore hard to digest. On days when they are sick, however, the same animals lose their caution and go after high-tannin leaves, perhaps because the tannins bind to and escort plant poisons out of their system. Red colobus monkeys (*Procolobus badius*) are shown to do the same out-of-the-ordinary selecting when nursing a stomachache.

What looks like a momentary flight from good judgment may actually be a trip to the tropical medicine chest, say the zoopharmacognosists. Of course, as is true of so much of animal behavior theory, no one can prove this—at this point it's a matter of conjecture and common sense. "Most other primatologists have been reluctant to accept this self-medication explanation," writes Glander in one of his papers, "but they have been unable to offer other cogent explanations for the occasional ingestion of tannin-rich plant material by primates such as red colobus and howlers."

In addition to physiological motivators, conditioning behaviors may also play a role in self-medication. The ultimate enforcer of bitter-leaf-eating behavior is to have a bitter leaf soothe an ailment. It's the flip side of the so-called Sauce Béarnaise syndrome, which causes an animal to associate negative body sensations with a particular food. Just as the scientist who named the syndrome is unlikely to order Béarnaise again, good experiences with a given food could have the opposite effect, acting to encourage that particular eating behavior.

Cultural learning may also help to shape the habit of self-medication. Bennett G. Galef, Jr., and Matthew Beck, psychologists at McMaster University in Ontario, observed that rats are more likely to try a cure for their ailments if they are surrounded by other rats that already prefer the food. Even if they've been conditioned to be phobic about it, they may give it a try if everyone else is doing it. We primates are especially good at mimicking behavior, which turns out to be a survival skill. The fellow who got sick on bad Béarnaise might have been spared if he had been able to watch his tablemate double over after eating the stuff. Similarly, if a chimp stumbled onto a good thing with the *Vernonia* pith, others would quickly see the wisdom in it.

As with smart eating, modeling Mom's behavior is probably the first way primates learn about medicinal plants. After they're grown, they watch and imitate how their troopmates handle illness. This sampling of good medicines may be another reinforcement for socialization. Says Kenneth Glander, "I think this is a social phenomenon to the extent that the group can contain much more knowledge than a single individual can, particularly when that knowledge is in a three-dimensional space—different leaves within the same tree have different properties and the material has to be handled in a different fashion when you are medicating."

While it's easy to conjecture how sickness might prompt an animal to treat itself, how do you explain the fact that perfectly healthy animals sometimes leave their troops and travel for miles to select certain plants at certain times of year? If the animals are not sick, what are they responding to? Sometimes the answer is easy. In the case of moose, a springtime gorging on aquatic plants is a quest for salt, which is largely absent from its winter diet. But what about animals that are not starved for nutrients, and yet spend their energy traveling to a particular plant at a particular time of year? Could they be preparing their body for something? Curious, anthropologist Karen Strier decided to accompany the muriqui monkeys of Brazil on one of their seasonal "food runs."

Plant Parenthood: It's Not Just for Stomachaches Anymore

To keep up with these beautiful monkeys, Karen Strier, author of *Faces in the Forest*, has to run pell-mell through Brazil's Atlantic Forest. Overhead, her subjects are like trapeze artists, swinging from branch to branch at breakneck speed. The males and the females grow to an identical size, the cap set by the need to be lightweight enough for branch-top travel. This equal stature helps make muriqui (*Brachyteles arachnoides*) one of the most peaceful and egalitarian species of primates ever studied. They are also, unfortunately, one of the world's rarest primate species. Habitat destruction has already claimed 95 percent of their home in the unique Atlantic Forest, and fewer than one thousand of the beautiful muriquis are left in a handful of isolated populations.

Keeping track of them in the remains of their jungle can be exhausting. Thankfully for Strier and her students, the monkeys take frequent breaks for feasting, mostly on fruit. When their mating season dawns, however, the muriquis suddenly switch horses. They ignore the fruits and set their sights almost exclusively on the leaves of two tree species in the legume family, *Apuleia leiocarpa* and *Platypodium elegans*. Upon analysis, Strier found that the leaves of both species are notably low in tannins, a substance known to interfere with protein digestion. Like Popeye squeezing open a can of spinach just before a fight, the monkeys may be looking for a surge of protein before mating, and therefore go for the more digestible, low-tannin leaves. The leaves may also contain compounds that prevent bacterial infections, which could help bolster the monkeys' health when they need it most.

Strier also noticed that besides eating different leaves, muriquis tend to take road trips during this time of year. They speed from the center of the jungle to the edge of their ranges, where the forest thins out into clearings. Here, they eat the fruit of a third species of legume, called *Enterolobium contortisiliquum*, or monkey ear. The fruit is full of stigmasterol, a phytoestrogen that we humans use to synthesize progesterone. Could it be, asks Strier, that the muriquis eat monkey ear in preparation for, or perhaps to influence the timing of, mating season? Is there such a thing as "reproductive eating"?

Kenneth Glander is asking the same question about mantled howler monkeys. He became suspicious when he recorded a number of highly gender-skewed births in howlers. Some females in the group were having broods consisting of nine out of ten males or four out of five females. This swamping of sexes cannot be understood by statistical averages.

Could it be, thought Glander, that the howler monkeys are eating something that might improve the odds of having either male or female offspring? Are they somehow changing the electrical environment of the vagina (by eating either acidic or alkaloid foods) and thereby either blocking or rolling out the red carpet for a particular sperm type? The idea is not so outlandish when you consider that a sperm carrying an X chromosome (female-producing) is electropositive, while a sperm carrying a Y chromosome (male-producing) is electronegative. Since like repels like, a negative environment in the vagina might block negatively charged sperm while assisting positively charged sperm. Glander tested his hypothesis by measuring the electric potential at the entrance to howlers' vaginas and at the cervix. There was enough of a difference in the millivolt readings between the two locations to convince him that, depending on what they ate, howlers might be able to "produce an electrical charge and change it from positive to negative."

If plants could be used to stack the gender deck, the plant-as-medicine theme expands to include plant-as-population-shaper. But why the manipulation? Glander explains: If the population is short on males, a female that produces males has a good chance of producing one that will be a troop leader. Producing a son who is a leader confers status on the mother (better access to food and safety, for instance). If the population is low on females, however, the mother may want to have females who will likely become first lady—making the mother an in-law of royalty. "All of us are familiar with the phrase 'You are what you eat,' " Glander says. "But I suggest that we may be what our mother eats."

Strier and Glander were not the first to postulate this phenomenon in mammals. In 1981, Patricia Berger found that plant compounds seem to influence reproduction in voles. If primates and even voles can influence when and if they will be fertile in response to environmental conditions, could it be that animals are in finer harmony with their environment than we have given them credit for?

At this point, we know of ten thousand secondary compounds, but chances are that animals, insects, birds, and lizards know of and have been experimenting with lots more. They may use them to prevent illness, to cure illness, maybe even to influence their fertility, abort their fetuses, or influence the gender of their offspring—all in response to environmental opportunities and limits of the moment. Compared to these real natives, we've been snooping around the jungle pharmacy for only a brief moment, long enough to know there's much, much more.

NOT MUCH TIME ON THE CLOCK

There was a time, not so very long ago, when we relied exclusively on plants, microbes, and animals for new drugs, and that's where we found 40 percent of all our prescription medicines. Here's a small sampling of what plants alone have given us in the field of pharmaceuticals:

- Taxol, isolated from the bark of the Pacific yew tree (*Taxus brevifolia*) in the Pacific Northwest, is a promising new drug used to treat ovarian and breast cancer patients.
- The steroid hormone diosgenin, isolated from wild yams (*Dioscorea composita*) in Mexico, was an essential ingredient in the first contraceptive pills.
- Vincristine and vinblastine, isolated from the Madagascar periwinkle (*Catharanthus roseus*), are used to treat Hodgkin's disease and certain kinds of childhood leukemia.
- A semi-synthetic derivative of the May apple (*Podophyllum peltatum*), a common woodland plant in the eastern United States, is used to treat testicular cancer and small-cell lung cancer.
- Digitalis, from the dried leaves of the purple foxglove (*Digitalis purpurea*), is used to treat congestive heart failure and other cardiac disorders.
- Reserpine, isolated from the roots of tropical shrubs in the genus *Rauwolfia*, is used as a sedative and to treat high blood pressure.

By the close of the 1970s, however, plants fell out of favor as candidates for pharmacological research. Soil bacteria and fungi kept yielding new antibiotics, and synthetic chemistry and molecular biology—under the rubric of "rational drug design"—were seen to be the next great source of drugs. We decided we didn't need plants to create our cures.

Today, conditions have conspired to bring plant sampling back in vogue. After a few decades of sifting through the soil in their own backyards, pharmaceutical companies are beginning to turn up the same old microbes, but no new drugs. Scientists are also finding it harder than they thought to synthesize drugs from scratch. Despite the billions of dollars spent in development, the long-awaited malaria drug, like many others, is stillborn in the lab. To compound matters, the FDA is cracking down on "me-too" drugs (existing formulas that, with a slight twist, can be sold under a different name). This prohibition makes it harder for drug companies to float financially while they wait for the next streptomycin.

In the meantime, disease is having no trouble holding up its end of the arms race. Epidemiologists say we are living in "the emerging age of viruses," battling new diseases like AIDS, while resistant strains of diseases that we thought we had under control, like tuberculosis and the bubonic plague, are back with a vengeance. Just when we need a breakthrough, we've reached a point of diminishing returns.

Once again, hopes are being pinned on nature's biochemical registry, which is billions of years in the making. "Given the high cost of chemical synthesis," says Charles McChesney, a natural products chemist at the University of Mississippi, "companies are increasingly inclined to let plants and other organisms do the synthetic work for them." In a flurry of exploration contracts, drug companies are heading outdoors to find their next big drug.

Between 1990 and 1993, five major drug companies joined the medicinal gold rush, announcing large-scale plans to prospect in seven countries. Most recently, the National Institutes of Health and several drug companies began a $2.5 million treasure hunt in the Great Barrier Reef off Australia, in Samoa, and in the rain forests of South America and Africa. In this effort, marine biologists and botanists will spend five years collecting approximately fifteen thousand marine organisms and twenty thousand plants. Meanwhile, a $2 million, three-year effort begun in 1993 with Pfizer, Inc., and the New York Botanical Garden will concentrate on plants here in the United

States. Also in the United States, the proposed Joint Program on Drug Discovery, Biodiversity Conservation, and Economic Growth would provide grants (funded by AID, NCI, and NSF) to develop drugs from the most promising plants. Meanwhile, a coalition of government agencies, nongovernmental organizations, and businesses both here and in Asia are collaborating to help local communities both use and preserve their forest and marine genetic resources. All told, a 1992 Office of Technology Assessment report listed some two hundred companies and nearly as many research institutions worldwide that are now looking for plants as sources of pharmaceuticals and pesticides.

Will this usher in a new era of resource plundering? Chemical ecologist Thomas Eisner from Cornell University doesn't think so. He believes that chemical prospecting can be essentially noninvasive both ecologically and culturally (as long as intellectual property rights are assigned to the local people—a system that was agreed upon at the 1992 United Nations Conference on Environment and Development in Rio). "Once biological activity is discovered," Eisner writes, "the usual procedure is not to harvest the source organism, but to identify the responsible chemical so it can be produced synthetically." For example, the natural opiates morphine and codeine were the *models* from which meperidine (Demerol), pentazocine (Talwin), and propoxyphene (Darvon) were then synthesized. Sampling for model design need not be extensive, says Eisner. In the case of "drugs from bugs," chemists need only small quantities for screening—about half a kilo of insects, or what hits several windshields on a tropical summer evening.

The last time many drug companies scoured the natural world for ideas was in the fifties. The jungles and reefs they will encounter in the nineties are very different—fragmented, fragile, and disappearing. Most frightening of all are reports that one in four wild species (includes all taxonomic categories) will be facing extinction by the year 2025. Underlying the new haste to find cures is the understanding that it may be now or never for chemical prospecting.

The job ahead is enormous. Out of the estimated 5 to 30 million living species on Earth (some estimates put it closer to 100 million) only about 1.4 million have been named. Less than 5 percent of the world's total roster of plant species have been identified, and out of the estimated 265,000 flowering species, only about 5,000, or 2 percent, have been studied exhaustively for chemical composition and medicinal value. To take one country as an example, scientists estimate that *nothing* is known about the chemistry of more than 99 *percent* of the plant species growing in Brazil.

To light a lamp in this darkness, companies and governments are pouring into the remaining pristine jungles and oceans, collecting examples of what is left. Back home, lab workers attempt the arduous task of analyzing the mountains of diversity on their warehouse floors.

CURE IN A HAYSTACK

Analysis is a complex process which attempts to separate the plant sample into smaller and smaller parts until the chemical of interest is isolated. The problem is that plants make so many compounds—up to five hundred or six hundred different compounds *in the same leaf*, each with fifty or sixty different biological activities. The real trick is to identify which one is performing the miracles.

First the sample is milled, distilled into a tarry sludge, and then treated with chemicals to separate out the essence of the plant. This essence is then pitted against many known human diseases to see whether it will take any action. The National Cancer Institute, for instance, is testing forty-five hundred samples a year, seeing if they have an effect on HIV-infected cells and sixty different kinds of tumor cell lines representing the various types of cancers, such as brain tumors, leukemias, and melanomas. (Ultimately the institute hopes to test twenty thousand substances a year against one hundred cell lines.) If a certain extract looks promising, it is further separated into its component chemicals, each of which is tested again. The most active ones are mapped on a molecular scale to see how their chemical structure may be contributing to what they do.

Once a promising molecule is identified, scientists can try to synthesize it in the lab, adding different twists in hopes of making it more effective. If a facsimile can't be made artificially, plant-tissue culture techniques may come to the rescue. A plant-tissue culture is a vat of plant cells all grown from a few starter cells. The cells produce copious amounts of the product, which is then separated out of solution. If the product passes all the tests of effectiveness, a company or government might try to invest the money needed to bring it to market—about $230 million for the average drug.

In the past, this assaying for bioactivity was a slow procedure—

you injected the extract into a rabbit and waited to see what happened. Bioassaying in the test tube has speeded up the process, but it's still a needle-in-a-haystack procedure. For every twelve thousand samples, only one becomes a drug, and its development (tweaking, enhancing, and testing the substance) can take ten years or more. In short, we're spending precious time in the lab screening unpromising compounds—and we don't have that kind of time. Experts agree we need to develop some sort of prescreening procedure to narrow our search and help us quickly key in on the promising compounds before the species that hold the recipes disappear.

How we have gone about narrowing that search tells a lot about us as a culture. At first, we simply dragged our collecting nets across the whole jungle in an indiscriminate approach. Collecting everything was easy, but the holdup was back at the lab—samples piled up waiting for analysis, and in the jungle, species were going extinct before we could even sample them. The fear was that by the time we found the cure for cancer or AIDS, then went back out to find more of the sample to study, it would be gone, plowed out by a bulldozer to make room for cattle or housing. There had to be a way to speed the search.

Next, we thought we'd be logical and try tracing the family tree of a sample that we found promising, hoping that related species would also contain powerful compounds. (For example, lilies are rich in alkaloids, so let's investigate the closely related orchid family. Bingo, they're rich in alkaloids, too.) This narrowing approach is called the phylogenetic strategy, but it's limited as well. Not all plant relatives arrive at the same chemical solutions to sticky predator problems.

Finally (and reluctantly), we in the Western world decided to officially solicit the help of shamans, indigenous folk healers from tribes that have been using the jungle pharmacy for centuries. We had relied on folk medicine heavily in the past, although this fact was never advertised. As Deputy Editor Philip H. Ableson writes in an April 1994 *Science* editorial: "Of the 121 clinically useful prescription drugs worldwide that are derived from higher plants, 74% of them came to the attention of pharmaceutical houses because of their use in traditional medicine." But we rarely advertised our sources, nor did we formally seek their help. Today, schoolchildren know the names Fleming and Pasteur and Salk, but the names of shamans in the Amazon and in Africa are on the tip of no one's tongue.

Finally, ethnobotany has begun to lose the stigma of a fringe discipline, and is now attracting both funding and professional personnel. A handful of organizations are trying to contact the last remaining indigenous cultures that have lived close to the Earth. Dialogues with their shamans have yielded several important compounds, including an oral hypoglycemic for diabetics, a respiratory virus fighter, and a possible antidote for herpes simplex. All three are reaching clinical trials thanks to a fleet-footed firm called Shaman Pharmaceuticals in south San Francisco that employs ten ethnobotanists on three continents. Another exciting prospect coming from folk remedies is prostratin, shown to be active in the test tube against HIV.

Fieldgoing ethnobotanists often speak of being outclassed by the native people, who have uncanny plant knowledge. The legendary Richard Evans Schultes, who has combed the Amazon for healing strategies for over forty years, writes that natives in the Amazon are able to differentiate between chemivars—plants that appear similar in form and yet have quite different chemical properties. Although Western botanists can't find any morphological differences among chemivars, the Indians can identify them by sight, even from many paces away. They say they base their identification not only on the physical look of the plant but also on its age, its size, and the kind of soil in which it grows. This sort of knowledge is dying, says Schultes, especially if healers do not have apprentices, or if their people have adopted pills over plants.

Cultural Survival, a culture advocacy group, estimates that the world has lost 90 of its 270 Indian cultures since 1900, about a tribe a year, and with them, all their knowledge. As Schultes writes in a March/April 1994 article in *The Sciences*, ". . . the Earth is losing not only the biodiversity of the forest; it is also losing what I call its crypto-diversity, the hidden chemical wealth of the plants." He calls on us to use native cultures as rapid-assessment teams already on the ground, but warns that as "civilization" encroaches, we can lose, in only one generation of acculturation, botanical knowledge acquired over millennia.

Ethnobotanists, then, like the biomimics, are also in a race. To narrow their search, they concentrate on cultures that are in floristically diverse areas, that transmit their healing knowledge through the generations, and that have resided in one place for long enough to explore and experiment with local vegetation. Based on those criteria, is there any culture that we're forgetting? Any source of local expert knowledge that we might be overlooking?

After spending time with Wrangham, Strier, and Glander, I immediately think of chimps and muriquis and howlers. All are local experts, passing knowledge from mother to offspring, and living in floristically diverse areas. Instead of thousands of years, these animals have been conducting millions of years of field trials. Their self-medication is more ancestral than that of indigenous peoples, and comes without the overlay of religious taboos or tribal customs. Why not let their "nose" for what is curative help us zero in on bioactive compounds, making the screening process more efficient?

Daniel Janzen, a tropical ecologist at the University of Pennsylvania, explains it this way: "I think there are better ways to spend money [than random sampling]—it's too broad shot. How do you know what to collect? How do you know which tree of the same species to collect? They differ in chemical composition—one tree may be stressed and one not stressed. Primates and birds and lizards know."

By getting to know them, say the zoopharmacognosists, we may begin to know, too.

ECOLOGICAL SLEUTHING: BIORATIONAL DRUG DISCOVERY

One of the most promising ways to explore the natural world, and to further narrow our search, is called biorational drug prospecting, a strategy advocated by Dan Janzen and Tom Eisner. The biorational route goes beyond simply following chimps and howlers around the jungles. It challenges us to use information from the *entire* ecosystem to find our target molecules. It requires that we know something about the relationships around us—the coevolutionary tangos of herbivore and herb, the community webs, the interlacing of population with bioregion. "I would use the whole set of animals out there," says Janzen. "Humans are just one animal . . . and they only pick the stuff that doesn't give them a stomachache or make them go blind." It is a detective game that dares us to use all our senses as well as our sense of ecology to ferret out the bio-clues.

Tom Eisner's father was a chemist who used to make cosmetics in his basement, leaving the lower floor "smelling in the most interesting ways." The younger Eisner developed an uncanny nose, allowing him to actually *smell* insects as he walked, identifying those full of potent chemicals. "Flying molecules," he calls them.

As an outgrowth of his work with insects, Eisner perfected the art of seeing what isn't obvious, finding, as he calls it, the "unforeseen from the unexpected." When prospecting for possible drugs, for instance, Eisner will look for plants that are notably *free* of damage. Plants that insects avoid eating must have potent defenses, he reasons, and should be screened for bioactive secondary compounds. Similarly, a tree that has no plant growth around its stem or is conspicuously free of disease should be checked for growth inhibitors or antibiotics that can serve as models for new herbicides and antimicrobials. If ants reject a fallen leaf, or predators avoid an insect's egg when it is covered with its mother's saliva, chemistry is at work, and ecology has handed us a clue.

Ecological sleuthing has already helped us zero in on compounds that naturally repel or kill insects. Commercially available formulations of natural plant-derived insecticides include nicotin, pyrethins, and rotenoids. These natural products are a welcome addition to a field of disturbingly less effective pesticides synthesized from petroleum. May Berenbaum of the University of Illinois at Urbana-Champaign describes the rat race we have entered with synthetic pesticides and the pests that learn to resist them. "The use of increasingly higher concentrations of existing insecticides has led to a fourfold increase in agricultural pests that manifest resistance to at least one type of insecticide. Where are the new pesticides? Not many have been developed since 1960, and the standing prescription is simply to spray more and more chemicals." A new crop of insecticides—those without a deadly residue that collects in animal tissues—would provide needed relief.

Other ways to find drugs from bugs is to watch how venomous animals handle both their enemies and their prey. Any substance that can have such a profound effect on the victim—paralyzing, poisoning, or even breaking down its cellular matter with a single dose—is bound to have powerful biochemical or pharmaceutical properties. Natural Product Sciences in Salt Lake City, with funding from the large pharmaceutical firm Pfizer, is looking into the toxins of spiders, snakes, and scorpions. These compounds, which attack specific neurochemical targets, are already helping researchers identify tiny openings in the membranes of human neurons that admit charged molecules called ions. Since ion channel activity is important in the signaling of nerve cells, the company hopes to develop drugs for relieving anxiety and depression, strokes, and degenerative neurological diseases.

Besides looking at individual organisms, biorational prospectors are also identifying *settings* that they believe will be particularly rich in toxins. Environments where animals must always be on their guard against high levels of disease or parasitism are like giant breeding grounds of chemical inventiveness. The defenses that animals evolve in these settings may yield magic shields for us as well.

The ocean tops the list of promising settings for biodiscovery, says D. John Faulkner, professor of marine chemistry at the Scripps Institution of Oceanography. Here, the sheer diversity of plants and animals far exceeds what you can find on land. Marine creatures are literally awash in the chemical byproducts of other creatures, and their watery world is teeming with microbes. To stave off poisons or diseases, they have had to defend themselves in novel ways.

A doctor named Michael Zasloff began to appreciate this when he noticed an extraordinary defensive immunity in dogfish sharks (*Squalus acanthias*); though they were often scarred in fights, they didn't develop infections. Looking closer, Zasloff isolated a powerful new antibiotic called squalamine from the shark. Zasloff also discovered—and was later able to synthesize—two slightly different strains of a powerful new antibiotic produced in frog's skin. The discovery grew out of his observation that surgical wounds in frogs healed without inflammation and were rarely infected after the frogs were thrown into a murky aquarium. The antibiotics, which Zasloff calls magainins (from the Hebrew for "shield") are the first chemical defense other than the immune system to be found in vertebrates. This doctor-turned-biomimic has since left his post as chief of human genetics at the Children's Hospital of Philadelphia to form Magainin Inc., a company founded on the idea of biorational drug discovery.

Zasloff isn't the only hunter at sea. C. M. Ireland of the University of Utah reports that during the 1980s alone, seventeen hundred compounds with bioactive properties were isolated from marine invertebrates. Despite this obvious wealth, it's only been in the last two decades that scientists have started systematically scouring the world's oceans for helpful chemicals.

As a general rule in biorational discovery, says Charles Arneson, of the Coral Reef Research Foundation, biologist-divers look for creatures that should be vulnerable, but aren't. For example, Spanish dancer (*Hexabranchus sanguineus*), a tasty-looking six-inch sea slug, is rarely bothered, despite the fact that it is not protected by a shell and moves at a slug's pace. Its secret shield turned out to be a noxious chemical that now forms the basis for an anti-inflammatory drug. Spanish dancer also puts out flowerlike egg masses that bio-chemist Faulkner says "look good enough to eat" but have no takers. Upon investigation, Faulkner and his students found that the sea slug sequesters powerful compounds from a sponge that it eats and con-centrates them in its eggs. These compounds do more than repel predators; they also have antifungal properties, and have shown some activity against human tumors!

Other examples of drugs from the deep that are being explored by U.S. scientists:

- Discodermolide, from the Bahamian sponge *Discodermia dis-soluta*, is a powerful immunosuppressive agent that may have a future role in suppressing organ rejection after transplant surgery.
- Bryostatin, from the West Coast bryozoan (moss animal) *Bug-ula neritina*, and didemnin B, from a Caribbean tunicate (sea squirt) of the genus *Trididemnum*, are both in clinical trials as cancer treatments.
- Pseudopterosin E, from the Caribbean gorgonian coral (*Pseu-dopterogorgia elisabethae*), and scalaradial, from dictyoceratid sponges found in the western Pacific, are both being studied as anti-inflammatory agents.

On land, biorational drug prospectors can find crowded, oceanlike conditions wherever colonies of organisms gather to breed in close quarters. Seals breeding by the thousands on the same beach, for instance, would provide a fertile environment for disease microbes to flourish, which would in turn encourage the evolution of micro-bial foes. Presumably, the individuals that managed to fight off in-fections in such crowded settings would be chock-full of ingenious antibiotics, some of which may benefit us.

Finally, we'd be smart to pay attention to "extremophiles"—creatures that survive scorching temperatures, months of being fro-zen, or extreme salinity. These tough cookies are the special forces of living creatures, thriving in environments that would wilt lesser species. By deliberately looking for creatures that awe us, we may just stumble upon a whole new chemistry—the spoils of survival.

SWALLOWING OUR HUBRIS

Not surprisingly, there is a great hue and cry against this biorational approach to drug prospecting. It seems that the idea of animals being wise about their world is something that is hard for some Sons and Daughters of Baconian Science to stomach. Robert M. Sapolsky, associate professor of biological science and neuroscience at Stanford University and author of *Why Zebras Don't Get Ulcers*, came down hard on the zoopharmacognosists in an opinion piece in the journal *The Sciences* in early 1994. He cautioned readers that the medicinal effects of pith drinking and leaf swallowing, which he called "eating episodes," were at the moment no more than anecdotal evidence. He claimed that zoopharmacognosists were treading into the New Age realm by attributing wisdom to animals without having the scientific evidence to back it up. How do we *know* the animal is actually getting medical doses from what it eats? asked Sapolsky. How can we tie cause to effect?

Wrangham, Huffman, Strier, and Rodriguez defended their new field in a subsequent article in *The Sciences*. They admitted that animal self-medication has not yet been *proven*, nor has it been shown that animals have innate knowledge of medicinal plants. They know there is a lot more work to do. But while Sapolsky says whoa, his targets say, look, just because it's complex doesn't mean we should abandon the possibility that there's some wisdom here that we could learn from. As the four zoopharmacognosists wrote in their rebuttal, "Even if this endeavor doesn't show us new drugs, we think it worthwhile merely because *there's a range of animal skills waiting to be discovered* [emphasis mine]. Of course it's tough when you depend not only on intimate knowledge of a population's behavior, but also on rare events that are difficult to manipulate experimentally. But we can take the anecdotes and build them into evidence."

Their closing remark could well become the rallying cry of biomimics everywhere: "In an era of shrinking biological resources, we don't think it's such a good idea to rely only on studies of laboratory rats. Let's keep an open mind about ways to explore the natural world." Amen.

Native Americans had no trouble accepting biomimicry. Long ago they acknowledged that they were led to their medicines by animals, most notably the bear. Tribes in Africa also reportedly turned to animals (their livestock) to find out what to eat after their drought-stricken crops had failed. As tribal leaders told Donald Vermeer, "We ate the plants they ate, found we were OK, and now we eat those plants." Even the U.S. Navy understands that animals may hold clues to our survival. In the U.S. Naval Institute's 1943 book *How to Survive on Land and Sea*, authors John and Frank Craighead write, "In general it is safe to try foods that you observe being eaten by birds and mammals. . . . Food eaten by rodents or by monkeys, baboons, bears, raccoons, and various other omnivorous animals usually will be safe for you to try."

Why has it taken so long for the rest of us to come around to what is so obvious—that animals living in this world for millions of years might lead us to foods and drugs? Perhaps it's the old specter of the belief that animals can't teach us anything. When I ask Kenneth Glander, he frowns beneath his handlebar mustache and says, "It probably has something to do with the fact that we think we're above animals. To say that we've learned something by watching a lower or nonhuman animal could be viewed as belittling to humans." He hears himself and stops. "See? Even the terminology—'lower animals' and 'nonhuman'—has a bias built in, and that bias is reflected in our reluctance to accept anything that is not human—and in some cases, even other humans' knowledge." He's right. We've only recently expanded our kinship circle to include indigenous cultures, to accept the so-called primitives' knowledge. It's taken those of us in the Western culture too long to do that, and in the process we've lost the opportunity to learn from tribes now scattered. Finally, we're beginning to include animals in our circle of consideration—hoping against hope that we are not too late.

For 99 percent of the time that humans have been on Earth, we watched the ways of animals to ensure our own survival as hunters and gatherers. Now, in a strange repeat of history, we are once again watching what they eat and what they avoid, what leaves they swallow whole or rub into their fur, and we are making notes to pass on to our tribe, the scientific community.

In some of the places we watch, the human connection to the Earth has been severed. There are no cooking fires to storytell around, no ceremonial dances to reenact the movement of the herds. Yet even in the most modern setting, there is indigenous knowledge in the collective wisdom of wild communities. Animals embody the same rootedness that made local people local experts—they are a living repository of habitat knowledge. This habitat knowledge gives animals the wherewithal to balance their diet, to incorporate new foods without poisoning themselves, to prevent and treat ailments, and perhaps even to influence their reproductive lives.

Wild plant eaters have already filtered and screened, assayed and applied the kaleidoscope of compounds that make up their world and ours. It is through them that we can tap the enormous potential of plant chemicals. By accepting their expertise, we may be retrieving the lost thread to a world we once knew well.

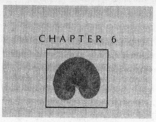

CHAPTER 6

HOW WILL WE STORE WHAT WE LEARN?

DANCES WITH MOLECULES: COMPUTING LIKE A CELL

> *Nerve cells are the mysterious butterflies of the soul, the beating of whose wings may someday—who knows?—clarify the secret of mental life.*
>
> —SANTIAGO RAMÓN y CAJAL, father of modern brain science

> *No one can possibly simulate you or me with a system that is less complex than you or me. The products that we produce may be viewed as a simulation, and while products can endure in ways that our bodies cannot, they can never capture the richness, complexity, or depth of purpose of their creator. Beethoven once remarked that the music he had written was nothing compared with the music he had heard.*
>
> —HEINZ PAGELS, author of *The Dreams of Reason*

I became curious about Jorge Luis Borges (1899–1986), the avant-garde Argentinean writer, after running across his quotes in so many of the mind/brain and computer books I was reading. His stories had made him a cult figure of sorts. When I read "The Library of Babel," I began to understand why. In that story, Borges asks us to imagine a huge library that contains all possible books, that is, each and every combination of letters, punctuation marks, and spaces in the English language.

Most of the books, of course, would be gibberish. But scattered throughout this vast library of possibilities would be books that made sense—all the books written, and all the books yet to be written. (At times, I agreed with Kevin Kelly, author of *Out of Control*, who wrote that it would be nice to visit Borges's library and simply find his next book without having to write it.) Surrounding these readable books, and fanning out in all directions in bookcases shaped like honeycombs, would be thousands of "almost books," books that were almost the same, except a word was transposed, a comma missing. The books closest to the real book would be only slightly changed, but as you got farther away, the books would degenerate into gibberish.

You could work your way up to a readable book in the following manner. Pick a book and browse it. Gibberish, gibberish, gibber— now wait a minute, here's one that has a whole word. You'd open a few more books, and if you found one that had two words and then three, you'd know you were on to something. The idea would be to walk in the direction of increasing order. If each book made more sense than the last, you would be getting warm. As long as you headed in the same direction, you would eventually come to the center of order—the complete book. Perhaps the book you are now holding in your hands.

Computer scientists call this library of all possible books a "space." You can talk about the space of all possible anythings. All possible comic books, all possible paintings, all possible conversations, all possible mathematical formulas. Evolution is like a hike through the "space" of all carbon-based life-forms, an upward climb past the contour lines of the "almost survived" to the mountain peak of survivors.

Engineering is also a form of bushwhacking through the space of all possible solutions to a problem, climbing toward better and better solutions until you reach the optimal peak. When we went looking for a machine that would represent, store, and manipulate information for us, we began the long trek toward modern-day computers.

What's humorous is that we forgot that we were not the only mountain climbers in the landscape of computing space. Information processing—computing—is the crux of all problem solving, whether it's done by us or by the banana slug on the log on which we are about to sit. Like us, the slug takes in information, processes it, and passes it along to initiate an action. As it begins oozing out of our way, our eyes take in the flicker of movement and pass it to our brain, saying, "Wait, don't sit." Both are forms of computing/problem solving, and evolution has been at it a lot longer than we have.

In fact, life has been wandering through the landscape of computing possibilities for 3.8 billion years. Life has a world of problems to solve—how to eat, survive capricious climates, find mates, escape from enemies, and more recently, choose the right stock in a fluctuating market. Deep inside multicellular organisms like ourselves, problem solving is occurring on a colossal scale. Embryonic cells are deciding to become liver cells, liver cells are deciding to release sugar, nerve cells are telling muscle cells to fire or be still, the immune system is deciding whether to zap a new foreign invader, and neurons are weighing incoming signals and churning out the message "Buy low, sell high." With mind-boggling precision, each cell manufactures nearly 200,000 different chemicals, hundreds at any one time. In technical terms, a highly distributed, massively parallel computer is hacking a living for each of us.

The problem is, we don't always recognize nature's computing styles because they are so different from our own. In the vast space of all possible computing styles, our engineers have climbed one particular mountain—that of digital silicon computing. We use a symbolic code of zeros and ones, processing in a linear sequence at great speeds. While we've been perfecting this one ascent, nature has already scaled numerous peaks in a whole different range.

Michael Conrad is one of the few people in computing who has stood on our silicon digital peak and taken a look around. Far off in the distance, he has spied nature's flags on other peaks and decided to climb toward them. Abandoning zeros and ones, Conrad is pursuing a totally new form of computing inspired by the lock-and-key interactions of proteins called enzymes. It's called jigsaw computing, and it uses shape and touch to literally "feel" its way to a solution. I decided to hike out to find him.

WHAT?! NO COMPUTER?

After reading Conrad's papers, I honestly didn't know whether to look for him in the department of mathematics, quantum physics, molecular biology, or evolutionary biology. He has worked for a time in all of these disciplines (I couldn't stop myself, he says), but these days, like a volunteer plant flourishing in a foreign ecosystem, Conrad brings his organic sensibilities to the most inorganic of sciences—computer science.

I was excited about going to see him. Although I make my home on the edge of the largest wilderness in the lower forty-eight states and adore all things biological, I am a shameless technophile when it comes to computers. I wrote my first book on a begged, borrowed, and all but stolen Osborn that had a blurry amber screen the size of an oscilloscope. I graduated from that to a sewing-machine-style Zenith luggable with a slightly larger green screen and the original, hieroglyphic WordStar program. I wrote the next three books peering into the monochrome scuba mask of a Macintosh SE/30 circa 1986 (a very good year in Apple's history). Finally, at the beginning of this book, I graduated to a Power Macintosh topped by a twenty-inch peacock of a monitor. I am completely smitten. To me, my computer is a semi-animated being, a connector to other inquiring minds on the Internet and a faithful recorder for every idea that stubs its toe on my receptors. In short, it's a mind amplifier, letting me leap tall buildings of imagination.

So, naturally, on my way over to Michael Conrad's office at Wayne State University in Detroit, I began to wonder what I'd see. Since he is head of the cutting-edge BioComputing Group, I thought he might be a beta tester for Apple and I'd get to see the next Powerbook or the operating system code-named Gershwin. Maybe he had a whole wall full of those flat-panel screens, controlled via a console/dashboard at his fingertips. Or maybe the desk itself would be a computer, ergonomic and wraparound, with a monitor built into eyeglasses and a keyboard that you wear like a glove. This would be something to see, I thought. Luckily, I had a few minutes alone in Conrad's office before he arrived—time to check out the gear.

It's strange. Here I am, in the lair of one of the most eminent minds in futuristic computing, and there isn't a CPU (central processing unit—the guts of the computer) in sight. No SIMMS, RAM, ROM or LANS, either. Instead, there are paintings. Not computer-generated laser prints, but heavy oils and watercolors with Conrad's signature on them. The largest, the size of a blackboard, looks like a fevered dream of the tropics spied through a lens that sees only greens, yellows, and blacks. It is disturbingly fecund—a hallucinogenic jungle of vines, heart-shaped leaves, and yellow blossoms, spiraling toward the viewer. A smaller painting of a painter—a Frenchman with beret and palette out by the docks somewhere—greets you as you come inside Conrad's office, as if to say, to visitors and to himself returning, the mathematician is really a painter. There are also paintings by his daughter. One on his desk has Picasso-esque double faces and daisy-petal legs going round in a wheel. I later find out that she is five, the age Conrad was when he asked his parents for oil paints.

Behind his desk sits an old Olympia typewriter (manual) and from the looks of fresh droppings of correction fluid, it still sees use. Finally I pick out the computer, nearly swallowed by a white whale of papers, journals, and notebooks. It's a yellowing Mac Plus from the early eighties, now considered an antique. When you turn it on, a little bell rings *Ta Da!*, and a computer with a happy face pops on the screen and says WELCOME TO MACINTOSH. I am perplexed.

When Conrad arrives, I recognize him from the French artist painting. He is without the palette but he does wear a maroon beret over a graying ponytail and zigzagging part. His eyes are so very alive that they almost tear up with emotion when he looks at you. He has caught me ogling his Mac Plus and he goes over to it. I expect him to throw an arm around it and tell me how important this machine was to the computer revolution. Instead he says, "This is the deadest thing in the universe."

PORT ME NOT: A COMPUTER IS NOT A GIANT BRAIN

In the forties, the term *computers* referred to people, specifically mathematicians hired by the defense department to calculate trajectories of armament. In the fifties, these bipedal computers were replaced by computing machines known colloquially as giant brains. It was a tempting metaphor, but it was far from true. We now know that computers are nothing like our brains, or even like the brains of slugs or hamsters. For one thing, our thinking parts are made of carbon, and computers' are made of silicon.

"There's a clear line in the sand between carbon and silicon," says Conrad, and when he realizes his pun (silicon *is* sand) he breaks into a fit of laughter that springs loose a few tears. (I like this guy.) He wipes his eyes and begins to paint a picture of the differences between the human brain and a computer, the reasons he thinks a silk purse will never be made from this silicon ear.

1. Brained beings can walk and chew gum and learn at the same time; silicon digital computers can't.
In the "space" of all possible problems, modern computers prove worthy steeds, doing a wonderful job of number crunching, data

manipulation, even graphic manipulation tasks. They can mix, match, and sort bits and bytes with aplomb. They can even make dinosaurs from the Jurassic Era seem to come alive on the screen. But finally, our steeds stall when we ask them to do things that we take for granted, things we do without thinking. Remember negotiating the crowded dance floor at your twenty-year reunion? Scanning a few feet ahead, you recognized faces from the past, put names to them, spotted someone approaching you, and recalling "the incident," you hid behind a tray of ham roll-ups. All in a split second. Ask a computer to do all this and you'd wait an ice age for a response.

The fact is that humans and many so-called "lower" animals do a great job of interacting with a complex environment; computers don't. We perceive situations, we recognize patterns quickly, and we learn, in real time, via hundreds of thousands of processors (neurons) working in parallel; computers don't. They've got keyboards and mice, which, as input devices go, can't hold a candle to ears, eyes, and taste buds.

Engineers know this, and they would love to build computers that are more like us. Instead of typing into them, we would simply show them things, or they would notice for themselves. They would be able to answer not just yes or no, but maybe. Spotting someone who looks familiar, they would venture a fuzzy guess as to the person's name, and if they were mobile (robotic), they would tap the person on the shoulder or wheel away, depending on what they had learned in the past. Like most of us, as our computers got older, they'd get wiser.

But at this point, all these tasks—pattern recognition, parallel processing, and learning—are stuck on the drawing boards. They are, in the words of computer theorists, "recalcitrant problems with combinatorial explosions," meaning that as the complexity of the problem grows (scanning a roomful of faces instead of just one), the amount of power and speed needed to crack the nut "explodes." The already blinding speed of modern processors can't touch the task. The question has become, how do we speed them up? Or more precisely, how do we speed them up if we're still stuck on controlling them?

2. Brains are unpredictable, but conventional computing is obsessed with control.

Today's computer chip is essentially a switching network—a railyard of switches and wires—with electrons (the basic particles of electricity) instead of trains traveling to and fro. Everything is controlled via switches—tiny gates at intervals along the wires that either block the flow of electrons or let them pass through. By applying a voltage to these gates, we can open or close them to represent zeros or ones. In short, we can control them.

One way to speed computers up would be to shorten electrons' commuting time by shrinking switches and packing them closer together. Knowing this, computer engineers have been "doing an Alice"—hanging out around the looking glass and itching to go smaller. Behind the mirror is a quantum world we can barely fathom, much less predict—a world of parallel universes, superposition principles, electron tunneling, and wayward thermal effects. As much as they'd like to cross that threshold, computer engineers acknowledge that there's a limit to how small electronic components can be. It's called Point One. Below .1 micron (the width of a DNA coil, or 1/500th the width of a human hair) electrons will laugh at a closed switch and tunnel right through. In a system built around control, this "jumping the tracks" would spell disaster.

Another route to speedier and more powerful computers would be to keep the components we have now but just add more of them; instead of one processor, we'd have thousands working in parallel to solve a problem. At first blush, parallelism sounds good. The drawback is that we can't be completely sure of what will happen when many programs are run concurrently. Programmers wouldn't be able to look in the user's manual to predict how programs would interact. Once again, control—the great idol of conventional computing—would do a faceplant.

When you look under the hood, you realize that we didn't build the "giant brain" in our image—we built it as a dependable, versatile appliance that we could control. The trick to predictable performance is conformity (as the military well knows). Standardized components must operate according to specs, so that any programmer in the world can consult the manual and write software that will control the computer's operations. This conformity comes at a price, however, which is why our computers, unlike our individualized brains, can't learn to learn.

3. Brains are not structurally programmable the way computers are.

In the silicon railyard of wires and switches, the modern-day switch-men are programmers. They write instructions in the special language of programming code, which we call software. When we double click on a screen icon, our software whirs to life, barking orders to the switches deep inside the computer, telling the gates when to open or close, connecting the tracks in new ways, and thereby changing the *structure* of the network, enabling it to perform a new function. Making the computer "structurally programmable" was the dream child of a man named John von Neumann. He wanted the computer to be the player piano of information—a universal device that could, with software to morph the network, become a word processor, a spreadsheet, or a game of Tetris.

Our brains, of course, are not structurally programmable. When we want to learn something, we don't read a book that tells us how to change our brain chemistry to remember a blues riff or the date of Delaware's statehood. We take on information, and our neuronal net is free to structurally store the data *on its own*, using whatever mechanical and quantum forces it can muster. Neuron connections are strengthened, axons grow dendrites, chemicals move in mysterious ways.

It's this physical processing, then, that makes our cells so different from our computers. While our PCs process information symbolically, with long strings of zeros and ones, our cells compute physically, working at the level of the molecule. We brain-owners take our lessons on an interpretive level—and the body *automatically* takes care of the rest. Michael Conrad's vision for computing is perched on this same peak.

4. Brains compute physically, not logically or symbolically.

Suddenly, Conrad holds his pencil high above his desk and lets go. "This," he says triumphantly as the pencil bounces, skitters, and rolls to a stop among his papers, "is how nature computes." Instead of switches, contends Conrad, nature computes with submicroscopic molecules that jigsaw together, literally falling to a solution.

Molecules are groups of atoms assembled according to the laws of physics into three-dimensional sculptures (think of the colorful ball-and-rod sculptures that scientists on *Nova* are always displaying). Large biomolecules can be made up of tens of thousands of atoms, and yet the finished object is still ten thousand times smaller than the cells in our bodies, a thousand times smaller than our silicon transistors. A molecule can't chip or erode, and though it can be bent or flattened, it'll always spring back to shape. The driving force at this scale is not gravity, but the push and pull of thermodynamic forces.

A molecule's goal in life is, like the pencil's, to fall to the minimum energy level—to relax. When two molecules free-floating in a liquid bump into one another so that their shapes correspond like jigsaw pieces and their electrical charges line up in register, there is an immediate attraction—an adding together of their weak forces—that is stronger than the urge to stay separated. In fact, it would take more energy at this point to keep them apart than to let them self-assemble. Like people falling asleep and finally rolling toward the sag in the bed, complementary molecules "snap" together as they relax. It's called "minimizing their free energy."

Right now, mix-and-match molecules are snapping together in every cell in every life-form on the planet. Conrad believes their fraternizing is a form of information processing, and that each cell in our brain, each neuron, is a tiny, bona fide computer. The brain manages to wire together *one hundred billion* of these computers in one massive network. (To get a feel for that number, come stand under the velvet Montana sky and check out the Milky Way. It's one hundred billion stars strong—one star for every person on Earth, times seventeen.) But there's more. Inside each neuron are tens of thousands of molecules engaged in a fantastic game of chemical tag set in motion each time, for instance, the phone rings.

It's 2:00 A.M., and you are in a hotel room fast asleep. The phone rings, setting off an amazing feat of computation, biology style. The first set of sound waves pounds like a hurricane against the hairlike cilia in your ear canal. These movements are turned into electrical impulses that wake you. Your body's mission is to integrate incoming signals, come to a conclusion, and do something, now.

Adrenaline molecules, the Green Berets of fear and anger, bail out of a gland and into your bloodstream, heading for nerve endings. At the shoreline of the nerve endings, molecules called receptors hold out their "arms" to catch the adrenaline molecules. Once the receptors are full, they change shape and "switch on" special enzymes inside the cell, which in turn activate a whole cascade of chemical reactions. The effects differ depending on the cell.

In your liver, the cascade may signal cells to start breaking down their stored sugar and swamping your bloodstream with glucose for fast energy. Your skin is told to tighten, your heart to speed, and your entire thirty-five feet of intestine to shut down (you have better things to do in a crisis than digest dinner). In your brain, the chemical cascade causes an electrical "action potential" to snake like a spark along a lipid (fat) fuse. At the end of its journey, it's not the spark that jumps from one neuron to another, but another boatload of chemicals. And it's this journey that most interests Michael Conrad.

The chemicals that are released from one neuron to another are called neurotransmitters (serotonin, the mood regulator affected by Prozac, is one example). These burst through the cell membrane at the end of one neuron and float by the hundreds across the liquid strait—the synaptic gap—to the shore of another neuron. Here they dock in the waving arms of receptors, which, in turn, change shape and trip off a series of their own chemical cascades deep inside the new neuron.

These chemical cascades cause gating proteins in the neuron's membrane to open, letting in a milling crowd of salt ions. This influx of charged particles causes the electrical environment of the membrane to reverse itself right at the point of entry. The outside membrane, which was once positively charged relative to the inside, becomes negatively charged relative to the inside in that spot. This flip-flop travels like an electrical shiver down the neuron, and at the end, it prompts the release of yet another barrage of neurotransmitters that float across the synapse to the next neuron. The result of all this is you remembering who you are and where you are and what a phone is, and picking it up just in time to become simultaneously furious (it's a prank) and relieved it wasn't something worse.

In crisis or in sleep, your body is busy at computational chores like this one. Carbon compounds in a million different forms are joining, separating, and rejoining to pass messages along. This process doesn't happen just in neurons, either—it occurs in less flashy cells as well. Shape-based computing is at the heart of hormone-receptor hookups, antigen-antibody matchups, genetic information transfer, and cell differentiation, just to name a few. Life uses the *shape* of chemicals to identify, to categorize, to deduce, and to decide what to do: how many endorphins to make for the blissful runner's high, which muscles to cause to contract, how many bacteria to kill, whether to become a tongue cell or an eye cell. Without shape-based computing, embryos—which begin life at the size of a period on this page and then divide only fifty times to become human babies—wouldn't be able to follow their recipe for development. We literally wouldn't be here without the chemical messenger system that is choreographed by shape-based, lock-and-key interactions.

When Conrad explains these "chemical cascades," he speaks as if he has floated across the straits of a synapse himself, ridden the fountain from the chemical signal up to the macroscopic electrical signal and back down to the chemical signal. "The most important conceptual journey for me was to go inside the neuron and slosh around at the chemical level," he says. "There, three-dimensional molecules are computing by touch. Pattern recognition is a physical process, a scanning process, not the logical process it is when our computers recognize a pattern of zeros and ones. Life doesn't number-crunch; life computes by *feeling* its way to a solution."

5. Brains are made of carbon, not silicon.
If you are going to rely on shape to feel your way to a solution, you have to use molecules that can assume millions of different shapes. Life knew what it was doing when it chose carbon as its substrate for computing. For one thing, carbon is free to participate in a great variety of strong bonds with other atoms and is quite stable once bonded, neither donating nor accepting electrons. Silicon, on the other hand, tends to be more fickle in its bonding, and is not able to form as many shapes as carbon can. As a result, Conrad believes life could not have evolved its shape-based computing using silicon. "And that's why, if we want to try *physical* computing as opposed to logical or symbolic computing, we have to eventually say goodbye to silicon and hello to carbon."

The clamor for carbon is not exactly heard across the land, however. Many artificial intelligence researchers are still putting all their faith in silicon. The sci-fi idea of "porting" our brains, or at least our thought patterns, to a computer host would supposedly allow us to live forever *in silico*. According to Conrad, it's the ultimate mind-body split. "It's absurd to think you can remove the logic of conscious thought from its material base and think you haven't lost anything. Even if you were able to put your thought patterns in a numerical code (the premise of 'strong' artificial intelligence theory), it would be only the map, not the territory. The territory, the seat of intelligence, is proteins and sugars and fats and nucleic acids—all carbon-based molecules."

Matter matters. And so, it seems, does the connectedness of this matter.

6. Brains compute in massive parallel; computers use linear processing.

Although neuroscientists have tried for decades to find the physical headquarters of consciousness, the grand central sage that organizes our thoughts, they have had to conclude that there is no central command. Instead, says author Kevin Kelly, the "wisdom of the net" presides. Thoughts arise from a meshwork of nodes (neurons) connected in democratic parallelism—thousands attached to thousands attached to thousands of neurons—all of which can be harnessed to solve a problem *in parallel*.

Computers, on the other hand, are *linear* processors; computing tasks are broken down into easily executed pieces, which queue up in an orderly fashion to be processed one at a time. All calculations have to funnel through this so-called "von Neumann bottleneck." Seers in the computing field bemoan the inefficiency of this setup; no matter how many fancy components you have under the hood, most of them are dormant at any given time. As Conrad says, "It's like having your toe be alive one minute, and then your forehead, and then your thumb. That's no way to run a body or a computer."

Linear processing also makes our computers vulnerable. If something blocks the bottleneck, that dreaded smoking bomb appears on the screen. The redundancy of net-hood, on the other hand, makes the brain unflappable—a few brain cells dying here and there won't sink the whole system (good news to those who survived the sixties). A net is also able to accommodate newcomers—when a new neuron or connection comes on line, its interaction with other neurons makes the whole stronger. Thanks to this flexibility, a brain can learn.

In an effort to imitate this brain-net in software form, a programming movement called "connectionism" has blossomed. In the last decade, "neural net" programs have been showing up on Wall Street, in manufacturing plants, and in political spin factories—wherever predictions need to be made. Neural nets are programs, like your word-processing program, that run on top of old-fashioned linear hardware. Inside your computer they create a virtual meshwork composed of input neurons, output neurons, and a level of hidden neurons in between, all copiously connected the way a brain might be.

Neural nets digest vast amounts of historical data, then seek relationships between that data and actual outcomes. At a campaign headquarters, for instance, a net might crunch all the polling and demographic data for 1992 and then try to find a relation between that and who won the New Hampshire primary. Eventually, you want your net to concoct a rule about it all, something like "If X and Y occur, then chances are Z will happen." Usually it takes some practice to come up with this rule, in the same way that a dog has to catch a few Frisbees before it makes up a rule about where a Frisbee will land. The neural net isn't a great predictor right out of the box; you have to train it by tossing it statistics from the past and having it guess the outcome.

Say a soda manufacturer wants a neural net to predict its sales figures in a particular town. It feeds the net reams of historical information: monthly temperatures, demographics, and advertising budget spent there in previous years. Given this constellation of conditions, the net connects its neurons in a certain way and tries to guess sales in previous years. At first, it ventures a wild guess. The trainer then feeds back the correct answer—the actual sales figures—and the net adjusts its connections and guesses again. It keeps readjusting its connections, revising its rule until it can correctly predict where the data will lead.

The reason nets learn so quickly is that connections between inputs can be weighted, as in, this input is more important than that input, so this connection should be strengthened. To the student of brain science, this theory of learning seems more than faintly familiar. In 1949, Canadian psychologist Donald O. Hebb postulated that memories (associative learning) were processed physically—the connections between neurons actually changed—and they grew stronger or weaker depending on whether neuron A had caused neuron B to fire. The idea was that the next time neuron A fired, neuron B would be more likely to fire because of some sort of "growth process or metabolic change" that strengthened the connection between the two. Hebb's guess was that dendritic, or branching, "spines" would grow between nerve cells to establish stronger connections. "It's the neurons-that-play-together-stay-together idea," says Conrad.

While our *in silico* neurons can't exactly grow spines, the network is able to adjust its connections again and again during a training process, all the while nudging toward a correct answer and, in the process, *embodying* a predictive model (a rule) in its network architecture. Once the winning network configuration is in place, these virtual neurons, run in virtual parallel, can quickly and uncannily reach the right solutions. In no time, they're catching the Frisbee on the run.

The next step, of course, is to build net-hood right into the hardware. Some computer designers are already etching neural nets onto silicon chips, while Thinking Machines, Inc., is hooking sixty-four thousand processors together into one giant Connection Machine. Assuming I could afford the $35 million model, I ask Conrad, would my new Connection Machine running a neural net be more like a brain?

"Connectionist hardware and software bring us closer," he says, "but they still miss an essential truth. Connections are important, but connecting *simple* switches or *simple* processors together is not how the brain got to where it is today." The brain astounds because every single neuron in the net is a wizard in its own right. And neurons are far from simple.

7. Neurons are sophisticated computers, not simple switches.

In the late sixties and early seventies, Conrad thought extensively about neurons and their interplay. "I began to realize that the neuron was a full-fledged chemical computer, processing information at a molecular level." His first papers about "enzymatic neurons" appeared in 1972 to somewhat skeptical reviews. "It's still controversial to call a neuron a chemical computer," he says, "but today, more and more neurophysiologists seem sympathetic to the idea. Finding someone who believed as I did twenty years ago—now *that* was a red-letter day.

"It was 1978 or '79, I think. A student came into my office and showed me an abstract of a paper on molecular computing by E. A. Liberman, and I thought, so there *is* someone else in the world using this term. I immediately arranged to visit his lab." Conrad spent the following year as a U.S. National Academy of Sciences Exchange Scientist to what was then the Soviet Union.

He and Liberman spent a lot of time talking about what makes neurons tick. Up to this point, neurons had been studied only for their response to electrical probings, the theory being that electrical impulses alone were responsible for thought. But as Liberman showed Conrad, neurons could fire without electric help. All a neuron needed was an injection of cyclic AMP, the chemical messenger that is instrumental in the cascade of signals leading to a neuron's firing. The shot of cAMP not only caused the neuron to fire, but "different concentrations of cAMP had the neuron talking differently and fairly rapidly to other neurons." It was a stunning sight, remembers Conrad.

Other labs were doing similar experiments. It soon became clear to other scientists that neuron communication was an electrochemical phenomenon, a dance far more complex than the simple "yes or no" of neuronal firing. When a neuron makes a decision, it has to consider some one thousand opinions coming from the axons attached to it. Instead of just averaging votes, it considers these opinions in detail. The receptors bobbing in the cell membrane are like doormen that receive messages from at least fifty different brands of neurotransmitter. The doormen in turn relay the message to "helpers" inside the cell who create secondary messages in the form of clouds of chemicals such as cAMP. Above a certain threshold concentration, cAMP turns on an enzyme called protein kinase, which in turn opens a gating protein. The gating protein causes a channel in the membrane to open or close, letting in or keeping out charged particles, thereby controlling the electrical shiver, and controlling whether and just how rapidly the neuron will fire.

To complicate matters, there is not just one doorman receiving the message, but several different doormen, all getting different messages, which they may or may not pass on to helpers. Inside, the helpers have their own conundrums. They may receive messages from more than one doorman, and must then decide which message to respond to. In certain cases, they may decide to combine the messages and respond to the net action of the two.

It's no wonder that Gerald D. Fischbach, chairman of the Department of Neurobiology at Harvard Medical School, agrees that the neuron is "a sophisticated computer." In a September 1992 article in *Scientific American* he writes: "To set the intensity (action potential frequency) of its output, each neuron must continually integrate up to 1,000 synaptic inputs, which do not add up in a simple linear manner. . . . The enzymes make a decision about whether the cells are going to fire and how they will fire. . . . [B]y fine-tuning their activity, [enzymes] may have an active role in learning. It may be *their* ability to change that gives us a malleable machine—the neuron."

Thinking is certainly not the yes-or-no, fire-or-not-fire proposition it was once believed to be. Each week, biological journals are filled with descriptions of newly discovered messenger molecules, helpers, and doormen. There's a cast of thousands in there, weighing and considering inputs, using quantum physics to scan other molecules, transducing signals and amplifying messages, and after all that computation, sending signals of their own. In silicon computing, we completely ignore this complexity, replacing neurons with simple on-or-off switches.

"When you want to find the real computer behind the curtain," says Conrad, "you have to put your cursor on the neuron and double click. That's where you'll find the computer of the future. What I want to do is replace a whole network of digital switches with *one* neuronlike processor that will do everything the network does and more. Then I'd like to connect lots of these neuronlike processors together and see what happens." By this point, I knew better than to ask him what that might be. When adaptable systems are involved, prediction is futile.

8. Brains are equipped to evolve by using side effects. Computers must freeze out all side effects.

"How is a brain like a box-spring mattress?" riddles Conrad. Answer: You take one spring out of a boxspring, and you're not likely to notice it because there are plenty of others. In the same way, nature builds in redundancy so that change, good or bad, can be accommodated. When we look at the nerve circuitry in a fish, for instance, we are appalled—it seems to be loops circling back on loops, as if nature's engineer was lazy, adding new circuitry without removing the old. Nevertheless, this seemingly messy system works beautifully. When part of it fails, other regions take up the slack.

Nature's redundancy is built into the shapely origamis called proteins too. Conrad draws me a schematic of a typical protein, a string of amino acids folded spontaneously into a lyrical but functional shape. He draws the amino acids as geometric shapes and connects them with either springs (representing weak bonds) or solid lines (representing stronger bonds). Having enough "springs" to accept change is the protein's secret to success. If a mutation adds an amino acid, for instance (Conrad draws in an exaggerated beach ball of a newcomer), the springy connections give to absorb the new player. This allows the active site—where chemical reactions occur—to remain undisturbed so it can continue to do its lock-and-key rendezvous. The fact that proteins can graciously accept incremental, mutational change without falling apart is important. It means they can improve over time.

Life experiments like a child at play, says German biophysicist

Helmut Tributsch. It dabbles in all the possible computing domains and learns to solve its problems creatively, harnessing every single force in the library of physical forces—electrical, thermal, chemical, photochemical, and quantum—to physically tune up neurons and their ways of communicating with one another. When small changes are permitted without a fuss, helpful effects gradually accumulate, and evolution pounces to a new level.

What would be a nightmare to computer engineers—quantumly small computing elements, connected catawampus in dizzying parallelism, randomly interacting and coloring outside the lines—is what gives life its unswerving advantage. If it needs to recognize a pattern, learn something new, or stretch to assimilate new information, it molds its substrate to the task, adding new elements, shaking up the works until it works. This is the world that biological organisms revel in. The ability to ride that riot of foreseeable and unforeseeable forces has allowed nature to exploit myriad effects, becoming more efficient and better equipped all the time. The power to be unpredictable and to try new approaches is what gives life the right stuff.

Our computers, by comparison, are in shackles.

Computers can't brook too much change. If you add a random line of code to a program, for instance, it's not called a new possibility—it's called a bug. Unlike biology, which built its empire on faults that turned to gold, computers can't tolerate so much as a comma out of place in their codes. Add a new piece of hardware to the inside of your computer, and no springs will adjust to accommodate it. The other components, which must remain true to their user-manual definitions, can't interact with the newcomer or take advantage of the new interactions to bootstrap themselves to anything more efficient. No fraternizing among the transistors; no conspiring or self-organizing allowed.

Unlike biology, which was able to transform the swim bladder in primitive fish into a lung, structurally programmable computers can't transform their function, hitch up additional horses, or get any better at computing. In essence, *they can't evolve or adapt*. When the really large problems crop up, they choke, and the bomb appears on the screen.

In the age of *Siliconus rex*, says Michael Conrad, "We feel powerful, but what we've really done is trade away our power for control. To make sure only one thing happens at a time, we've frozen out all interactions and side effects, even those that could be beneficial or brilliant. As a result we have a machine that is thoroughly dead—inefficient, inflexible, and doomed by the limits of Newtonian physics.''

And I had thought he was going to throw his arm around that old Mac Plus and gush.

The nice thing about articulating the differences between brains and computers is that it gives you a clear mandate: If you want better computers, better stay to the brain side of the chart. First, design processors that are powerful in their own right. Fashion them in nature's image by using a material that's amenable to evolution, embedded in a system with a lot of springs. Then, when you challenge your computer with a difficult problem, it'll hitch all its horses to the problem. Efficiency will soar. And when conditions change, and it needs to switch horses, it can adapt.

So when Michael Conrad, way back in the seventies, went looking for a new computing platform, he had one big item on his wish list. He didn't care if it was fast, he didn't care if it could compute pi to the infinite decimal place. He didn't even care if it could sing and dance. "I just wanted it to be a good evolver."

JIGSAW COMPUTING

Back in those days, Conrad was thinking quite a bit about evolution at the molecular level. "I was in an origin of life lab and my professor wanted me to model the conditions necessary for evolution to evolve. I was to create a world in silicon, using linear string processing to represent proto-organisms that would have genotypes, phenotypes, material cycles, and environments—they would eat, compete, die, mutate, and have offspring. I was to find out what conditions would foment evolution and encourage the players to bootstrap themselves to higher states of complexity."

Conrad eventually created a program called EVOLVE—the first attempt at what is now called artificial life. "If I had claimed it was artificial life," he says, "those programs would be more famous than they are today. But I didn't see it as life; I saw it as a map in action." Nevertheless, the exercise bore fruit and seeded his dream of nature-based computing. He says it happened one night when a dog was barking and he couldn't sleep.

"I lay awake in active thought for hours. I was resisting the idea of using string processing language because it wouldn't allow me to capture the essence of biological processes. Biological systems don't work with strings, I realized; *they work with three-dimensional shapes.*"

In nature, shape is synonymous with function. Proteins start out as strings of amino acids or nucleotides, but they don't stay that way for long. They fold up in very specific ways. To put it in computing terms, it would be like putting Pascal programming language on magnetized beads. The program would run by folding up into a fork or spoon, thus determining its function—whether it could be used to stab a steak or slurp up bisque.

Because molecules have a specific shape that can feel for other shapes, they are the ultimate pattern recognizers. And pattern recognition is what computing is all about! Patterns are not just physical arrangements in space, they can also be symbols—the Morse code is a pattern language, for instance, as is binary mathematics. Computing works because each switch in the tiny railyard recognizes a pattern of zeros and ones.

Conrad began to fantasize. What if we built processors full of molecules that recognized patterns through shape-fitting—lining up like corresponding pieces of a puzzle and then falling together, crystallizing an answer? In this way, he thought, a lovely irony could occur. The pattern recognition that tiny molecules are so good at could be hitched together by the millions and used to solve larger problems of pattern recognition—like recognizing a face in real time in a complex environment. Acting as the Seeing Eye dog for digital computers would be a natural job for the efficient, parallel, and adaptable shape processor. And that would be only the beginning.

"As I lay there I realized that the world's best pattern processor, a protein, is also amenable to evolution. If we used proteinlike molecules to compute, we could vary them, or rather, allow them to mutate, tweaking their own amino acid structures until they were fit for a new task. Here was my evolver! In a rush, in a vision, the 'tactilizing processor' came to me."

Science writer David Freeman calls the tactilizing processor a computer in a jar, although there's no saying what physical form it might take—it could float in a vial of water, or be trapped inside a hydrogel-liquid wafer as thin as a contact lens. Whatever form it took, the surface would no doubt bristle with receptor molecules— sensors—that are sensitive to light. Each receptor, when excited by a different frequency of light, would release a shape (a molecule) into a liquid. One receptor might release a triangle, the other a square, the third a shape that would join a triangle and a square. These released molecules would then free-fall through the solution until they met the shapes that complemented them. These three shapes would dock together jigsaw style into a larger piece—a "mosaic"—that would geometrically represent the incoming frequencies, the light signals. Different mosaics would be a way of categorizing the light inputs, or naming them.

Let's take an example. An image of a snowshoe hare is flashed onto the membrane surface (actually the image would be projected at a whole array of processors, but we'll keep it simple). The excited receptors release their shapes, and each shape represents a part of the image—long white ears, big feet, whiskers. The self-assembled mosaic of those shapes says "snowshoe hare." This naming, or generalizing from specific inputs into a category, is what our vision system does all the time.

Say you walk into a strange room and you see a chair you've never seen. It could be a kitchen chair, or an office chair, or an art sculpture of a chair covered with hair, yet your brain pegs it as a chair. It sees a place to sit down, a back, and four legs and shouts "I know, I know! It's a chair!" Coding is also how the immune system works. When an immune cell recognizes a certain concentration of foreign objects on its membrane, it integrates those signals into a category—"We have a particular disease problem"—and it begins manufacturing the antibodies needed to fight the disease.

For proof of his coding theory, Conrad points to the relatively small number of second messengers inside the cell compared to the vast number of messages impinging on the cell. "The fact that the cell employs so few second messengers to transduce [translate] this deluge of information is telling," says Conrad. "It shows that there must be some kind of coding, or signal representation, going on in the cell."

In the tactilizing processor, the mosaic will play the role of the secondary messenger, transducing the signal and posting the answer in the form of a unique shape. Just as a cloud of cAMP in the neuron says "serotonin has arrived," the mosaic's shape will say "snowshoe hare." But since the snowshoe hare mosaic is molecular (too small to be seen with the naked eye), we humans will need a way to amplify and read out the result of the computation. In the neuron, an enzyme called protein kinase "reads" the concentration of cAMP and responds to a threshold amount by opening or closing channel proteins. The enzyme in Conrad's tactilizing processor will read the "snowshoe hare" mosaic by touch, and instead of opening a channel, it will get busy churning out a product that we can measure.

The activated enzyme may grab two substrates in the solution, say chemical A and chemical B. Like a little machine, it will join these into product AB, then grab some more. After a time, the concentration of AB increases to the point that its characteristics can be measured by something like an ion-sensitive electrode or a dye that changes color when the pH or voltage changes. In this way, the enzyme amplifies the invisible to the visible.

Amplification schemes like these are used in biosensors all the time. In at-home pregnancy or cholesterol tests, for instance, receptors are immobilized on the surface of the tester, and when their open arms "catch" telltale molecules in your blood or urine, the receptors change shape. This shape change cues an enzyme to do its thing, usually a chemical reaction. Suddenly, as you stare at it, the stick turns blue.

In the tactilizing processor, the inputs would be light signals, and the "stick" would actually be a whole array of light-receptive processors. Each processor would recognize a bit of ear, a bit of tail, and so on, and when they were combined, the entire image would be recognized. Without a single electric wire or silicon circuit, a large number of disparate signals would be sorted, coded, and translated, simultaneously, into a coherent answer.

Given the time it takes for objects to float through liquid, however, is jigsaw computing fast? "No, actually. It's not," says Conrad. "Compared to a digital switch, the action of a readout enzyme would be up to five orders of magnitude slower." This doesn't seem to worry him, however. "Remember that we are not trying to do what silicon computers do well—we're not hoping to beat them at their own game." Digital computers, with their ability to perform repetitive operations at great speed, are perfect at recognizing bar codes and typewritten characters because the domain—all possible typewritten characters and stripes—can be whittled down to something finite that you can place in the computer's memory banks. But when you open up the domain to anything and everything that might hop past the sensors, you need a lot more than speed.

The advantage of scanning shape to arrive at a conclusion is that you are able to consider all the inputs—they all contribute to the shape-matching process, so each is fully represented in the final conglomerate, the mosaic. By contrast, silicon terminals simply *average* the inputs of zeros and ones to decide whether to let electrons through or not. This averaging actually blurs the inputs. If you were to force a conventional computer to be more precise—to in fact replicate the thorough scanning that floating molecules do for free, it would take our most powerful computers thousands of years. Conrad politely calls it "computationally expensive" and doubts whether it's possible at all.

Besides, he says, tactilizing is not as slow as it seems, thanks to quantum mechanics. Conrad's latest articles are all about the "speedup effect," which may explain why molecules snap together faster than predicted at normal Brownian mingling rates. He thinks that electrons are constantly "trying out" all possible orbitals or energy states, searching for the minimum, the spot where they can relax. Because of a quantum phenomenon known as quantum parallelism, they can actually explore more than one spot at once in the energy landscape. This parallel scanning allows two molecules to quickly line themselves into register and snap together for a secure fit. Our computers, with their strictly controlled regimes, couldn't possibly be in two places at once. They might be able to digitally find a minimum energy level, but they would have to go through each and every possible conformation, one at a time. A glacially slow proposition.

Another plus for the computer in a jar is its inborn talent for fuzzy computing. Patterns may dribble into the receptors, distorted in space or time, but the shapes floating in the medium will still find one another and compute the right answer. Given the flexible nature of shapes, mosaics, and enzymes, a good guess is likely to crystallize even if the inputs are faint or garbled.

To everyone's amazement, computing in this most natural of ways, going with the flow of physics and away from absolute control, turns out to be the most powerful form of computing. It's both precise and fuzzy, depending on what is needed, and it handles vast oceans of data with ease.

The question remains: When will tactilizing processors be sloshing inside in my Powerbook? Conrad, beret and all, is a pragmatist. He has a good feel for the biocomputing field, having been the elected president of the International Society for Molecular Electronics and Biocomputing for a number of years and serving as an editor and board member on several international computing journals. "In one of our very first conferences," he remembers, "I was thrown into a piranha tank of news reporters who heard we were building organic computers. They wanted to know *when*. Being very generous, I said fifty years, and their faces fell."

What Conrad means is that we'd need at least fifty years (I wanted to say a thousand, he admits) to have a computer built on shape-based principles only—which for him is the best of all possible worlds. Between now and then, however, you are likely to see more and more hybrids cropping up—conventional computers with organic prostheses attached. For example, his tactilizing processor may be the eyes and ears—the input device that predigests ambiguous information and feeds it to the digital computer. Tactilizing processors might also show up at the output end of things, as actuators—the devices that move the arms and legs of robots. While each tactilizing processor would be a computer in its own right, they would be small enough to be hooked up in parallel, perhaps connected in neural network designs. This team of complex processors would be more powerful, and more task-specialized, than anything we work with today.

We have miles to go before realizing even this halfway dream, however. As Felix Hong, a coworker of Michael Conrad's, emphasizes, "There is no infrastructure in molecular electronics as yet. You can't go to a catalog and order parts to make a computer like this. Biosensors are the closest thing we have, and we would no doubt build off that technology for receptor and readout parts of the processor. But everything else—the macromolecules, the system design, the software—it all has to be made from scratch."

And that's where breeding comes in.

Computer, Assemble Thyself

There may not be a catalog of molecular computer parts, but in Conrad's head there exists a factory, which he describes in papers as the molecular computer factory. It's unlike any factory we've ever seen, he assures me. It's more like a giant breeding facility, mimicking nature's tricks of evolution. Each element, both hardware and software, will be bred, through artificial selection, to do the best possible job and to interact well with other parts of the system. In this coevolutionary way, the molecular computer factory will resemble an ecosystem made up of different "members" challenging each other to work seamlessly together and up the ante of performance.

Conrad describes it this way: "Instead of being controlled from the outside, by us, each processor will mold itself to the task at hand, while together, several processors will sharpen their ability to work as a team. They will actually evolve through a process of variation and selection toward an optimal peak, the best possible system for the conditions at hand.

"We as engineers will coach the process. We'll be the invisible hand of natural selection, winnowing out the losers and putting the winners through increasingly tougher trials. Our biggest challenge won't be to create solutions (those will be generated randomly, the way species' adaptations are), but rather to describe the task we want done and then set up the evolutionary criteria—the environment that challenges the evolving forms to do their best. This is a whole new way for engineers to think."

It may be new to computer engineers, but stepping into nature's shoes and "defining the evolutionary criteria" is something with which we humans are very familiar. Ten thousand years ago our ancestors started to get choosy about the plants they ate and began saving the seeds of the tastiest, best-germinating, most uniform plants, tossing the rest over the garden gate. We were showing gene favoritism way back then.

Today, we have the awesome (and somewhat frightening) power to isolate our favorite genes and make millions of copies of them. We can insert a gene that produces insulin, for instance, into bacteria and essentially borrow their protein synthesis machinery to make insulin for us. Conrad would use a similar scheme, but instead of insulin, he would want the *E. coli* to produce jigsawing macro-molecules, light-sensitive receptors, and readout enzymes. The DNA blueprints for these molecules would probably be synthesized from scratch on oligio machines (which string together DNA bases into strands).

"Finding the best structure for these molecules will be an evolutionary process," says Conrad. "We'll let the molecules, receptors, and enzymes strut their stuff in tactilizing processors, seeing how well they can recognize a test image. Each time they make an error, we'll break apart the mosaic and let them try a new configuration. Just as biological systems are adept at finding a steady state, so too will the computer in a jar settle into a workable scheme for computing.

"Swarms of variation trials would be running simultaneously with various teams of processors being played off against each other to see which one solves a problem most effectively. Each trial will yield star performers, who, like prize pigs, will be bred again for the next trial. We'll encourage a mutation here and there, and then let them compete against their peers. Eventually, after a surprisingly small number of trials (thanks to the cumulative improvement power of variation and selection), we'll have our custom-designed team."

Though it sounds outrageous at first, this idea of "directed evolution" has already proven worth its salt in the medical field. Gerald Joyce of the Scripps Research Institution in La Jolla, California, got everyone's attention in 1990 when he announced that he was letting drugs design themselves.

The technique is deceptively simple. Drug manufacturers often know that they need a molecule with a certain shape that will interfere with a disease mechanism—by clogging a receptor, for instance. Instead of designing it by hand, they mutate a starting molecule to produce billions of variants. They test those molecules by floating them past billions of receptors. The molecules that dock even partially are kept for the next trial. These are copied, mutated again, tested, and culled again. Since the fit keeps getting better and better, Joyce found that he was able to manufacture his first product (an RNA molecule called a ribozyme that cuts DNA in a specific place) in only ten generations. Now directed evolution, the biommimicking of natural selection, is being pursued by dozens of companies.

Survival of the Fittest Code

OK, I tell Conrad, test-tube evolution is a long way from the pea gardens of Gregor Mendel (the monk who first fathomed the rules of heredity), but at least the molecules inside are biological. I can imagine natural selection working its magic on them, because they are organic and three-dimensional. But how do you plan to breed system designs, neural-net architectures, and software programs, all of which live exclusively *in silico*? How does one go about breeding strings of information, or programming code?

Computers, as it turns out, are dandy breeding devices. Say you are an artist, and you want to evolve art on the computer. You write a line of programming code that will instruct the computer to draw a pyramid and then you tell the computer to slightly mutate this pyramid. You run the program twenty times and get twenty different pyramid variations. You then use your aesthetic sense to pick an attractive variant that you will allow to survive. You have this survivor's DNA (the programming code) copy itself with further mutations and draw out twenty new variations and pick another winner. Choose again, and again, and again. Each choice is nudging the drawing toward the artist's ideal form—as if the artist is climbing the landscape of all possible forms to find the final, fully evolved form. This is already happening in a worldwide experiment called evolvable art on the World Wide Web. People vote for their favorites, and the group's choice of code is then used to redraw the pictures, with slight mutations, every thirty minutes.

In 1985, Richard Dawkins, zoologist and author of *The Blind Watchmaker*, took a similar journey of exploration inside a computer. Instead of art pieces, he was investigating biological forms. He was looking for the common denominators among biological forms, and so he wrote a program that gave the computer instructions for drawing a form. The instructions were simple rules, such as "draw a 1-inch line, fork it into two 1-inch lines, and repeat." He then gave the program parameters such as "maintain left-right symmetry."

In all his years of crawling around jungles as a zoologist, Dawkins says he never experienced anything quite like the rapid blossoming of forms in his computer. Starting from complete randomness, his program managed to make something that looked vaguely biological within a few generations. When it did, Dawkins chose the most biological-looking forebear and had the program begin here, modifying this form. At each stage, he chose forms that looked more and more biological, until he began to recognize forms that actually exist in nature. That night, as the computer drew tulips and daisies and irises, he couldn't pull himself away from the machine to eat or sleep.

Early the next morning he decided to step back and start in a new direction with his selection. Amazingly, the program yielded beetles and water spiders and fleas—he'd run into the domain of insect forms! Instantly, Dawkins saw parallels between the instructions in his program code and genes. It was as if his programs were genes that, once "run," came out with a phenotype—a drawing. Changing the instructions in the program was like changing genes to produce a slightly different individual. It was variation, which, when combined with selection of a winning offspring, was the formula for evolution.

What a powerful method this artificial evolution is for finding an optimum solution! What if instead of an insect or a tulip drawing, you used artificial evolution to design a jet aircraft? You could give the computer some criteria—weight, cost, materials, say—and let it begin to spin out a program code for the design of a jet aircraft. That code could be copied faithfully and it could be copied with mutations. As John Holland, the father of genetic algorithms, found, you can even have your program codes undergo mating. To "mate" two programs, you join half of one program's string of code to half of another program's string. The offspring is a thereby a mix of the two "parents." With this digital sex, the generations of programs literally fly by, pausing only for testing against criteria that you select. Design programs that meet these criteria are mated to produce even better designs, which are once again tested. The selection process heads in one direction—successful designs survive and suboptimal ones "die" out of the population. This "hill climbing" in a landscape of possibilities toward an optimal design is what engineers do, but computers can generate random ideas much faster than most engineers. And computers, not yet able to feel embarrassment or peer pressure, are not afraid to try off-the-wall ideas. Ideas are just ideas; the more the merrier.

Giving Up Control

As computing tasks become more complex—running a telephone system, flying a space shuttle, delivering electricity to more homes—our systems become harder to centrally control and repair. If we are to break out of our control-hungry straitjacket and achieve true power, says Conrad, we may have to loosen up the reins a bit. We may have to give computers their head, so to speak, give them the substrate (carbon) and the computing environment (artificial evolution) they need to creatively problem-solve so they can avoid troubles and perhaps even repair themselves. In the ultimate molecular computer factory of Conrad's imagination, self-improvement regimes will be built into the computers, so that when they run into snags, they'll be prompted to "create a new program using artificial evolution" until operations are smooth once again. Instead of crashing, they'll adapt to changing conditions without having to go offline for repairs.

What's hard for some to accept is that we're not the ones coming up with the solutions, and we may not wholly understand *why* they work as well as they do. Michael Conrad isn't a bit bothered. "I knew that I would have to give up control if I hoped to get real power, which is the power to adapt. I may not know where every single electron is, and I may not know why my molecular, shape-based device is doing such a good job. I'll just have to evolve it, test it, and marvel at how well it works without knowing exactly why."

This is the essence of the "letting go" that Conrad talks about. It is counterintuitive to the engineer who was schooled in the old way—being graded not only on the solution but also on how he or she derived the solution. This new paradigm asks us to admit that some approaches may work or even be superior to our own, even if we don't recognize them as something that would have sprung from our imaginations.

Life is like a rodeo—you can fight the bull's every buck and be worn to a frazzle (if you aren't gored first), or you can match your movements to your mount and see where it takes you. Deep inside our cells, where all the computing is going on, it's still the Wild West. Proteins tumble in a maelstrom of Brownian motion, riding a riot of electrical attractions, quantum forces, and thermodynamic imperatives. The computer networks that can match their movements to these forces, says Conrad, are going to astonish and, sometimes, humble us, as only carbon-based creations can.

SILICON COMPUTING IN A CARBON KEY

Computing is not liable to convert to carbon overnight, however. Conrad acknowledges that we have an enormous investment in the silicon-based computers sitting on our desktops. Most of our data is now encoded in zeros and ones. One way to begin the transition to the biocomputer is to practice a hybrid of silicon and carbon computing—keeping the on-off switches from the silicon past, but replacing the silicon with molecules from nature.

Conrad calls it "silicon computing in a carbon key." It doesn't change the fundamental approach to computing—that remains digital and linear—but it does bring organic molecules into play. Conrad doesn't say as much, but I get the feeling he thinks using biomolecules to crunch zeros and ones is like using a Lamborghini to deliver newspapers. He'd rather put natural molecules through their real paces by utilizing their shape-matching talents, but, he concedes, it would be kind of fun to capitalize on their light-reacting capabilities right now.

These days, one of the most promising avenues for speeding up computers is to think about abandoning electrons and using light pulses to represent zeros and ones. Many biological molecules are highly reactive to light. Some proteins actually move in predictable ways (they kink and unkink) when hit by certain frequencies of light. These proteins can be embedded in a solid material at densities orders of magnitude higher than conventional switches, and can be turned on and off via light waves—no tunneling electrons to worry about, and no buildup of heat.

It sounded like a peak in the computing landscape worth visiting. At Michael Conrad's suggestion, I contacted one of molecular computing's gurus, a man who, according to Conrad, knows everything you've ever wanted to know about kinking proteins but were afraid to ask.

When Light Flips the Switch

Felix Hong is an irrepressible host. At 9:30 P.M., the lab is empty, and he's unwrapping a new set of mugs. "Green tea?" Time slides by on stockinged feet when you are talking about someone's favorite molecule, and bacteriorhodopsin (or, as its friends say, BR) is Hong's very favorite. In the wild, BR is found spanning the membrane of a tiny, rod-shaped, flagellum-wielding bacterium called *Halobacterium halobium*. *Halobacterium* and its clan have survived for billions of years, in no small measure because of this strange protein in its cellular "skin." In a poetic turnabout, this most ancient of proteins is now one of the hottest stars of molecular electronics, poised to fill a new niche in sixth-generation computers.

Next time you fly into San Francisco, Hong tells me, look for the purplish smudge at the southeastern end of the Bay (toward Silicon Valley). That's *Halobacterium* by the billions, living, reproducing, and fighting for survival in some of the harshest conditions life can handle. The daytime temperatures soar, the nights are cold, and the water is ten times saltier than the Pacific—enough to pickle most creatures. "Salty is a relative term," he reminds me. "*Halobacterium*'s other favorite haunt is the Dead Sea."

These days, many laboratories around the world are trying to make *Halobacterium* feel at home. Engineers are growing the supertolerant microbe in bulk, hoping it will be a willing ally for enzyme and bioplastics manufacture, desalination, enhanced oil recovery, and even cancer-drug screening. Besides being tough to kill (even at 100 degrees Celsius), it's also full of strange engineering firsts, a brilliance born of adversity.

For one thing, *Halobacterium* can toggle from being a food consumer to being a food producer. When conditions are good, explains Hong, it gathers food that other creatures produce, and metabolizes it, just as we do. But sometimes, when oxygen levels in their shallow sea home dip and there is no way to oxidize, or burn up, food, *Halobacterium* goes to Plan B. It assembles in its membrane a protein called BR that allows it to harness sunlight to make its own sugars.

"Let me tell you how we think this works," Hong says, launching into a summary that is the distilled liquor of thousands of studies (two hundred papers a year have been published on this one molecule since it was first discovered in the seventies). Basically, sunlight causes BR to change its shape in the membrane. As it moves, it hands a proton—a positively charged hydrogen ion—from the inside of the membrane to the outside. Photon after photon pumps proton after proton, until eventually there's a buildup of positive charges outside the membrane relative to inside—a membrane potential that is poised to do work.

The protons on the outside of the membrane are like water in a high lake that wants to get back to the valley, to restore the balance of energies. Their only way back into the cell is through the "turbines" of ATP synthase, another molecular machine that spans the membrane. As the protons move through this tiny turbine and back into the cell, ATP synthase extracts a toll; it uses the energy to attach a third phosphate to adenosine diphosphate, making adenosine triphosphate, or ATP. ATP is then a molecular cache of energy—when the bacterium needs a boost, it can sever the high-energy phosphate bond, breaking ATP down to ADP, and releasing the energy that came originally from the sun.

"So you see," says Hong, with admiration lighting his face, "BR is both a photon harvester *and* a proton pump. It is also a smart material—whereas most pumps would slow down due to the 'back-pressure' of protons on the outside of the membrane, it adjusts to keep pumping protons. We admirers of this intelligent molecule are like corporate spies trying to reverse-engineer a machine that is only fifty angstroms by fifty angstroms, or one five millionth of an inch long."

After rummaging through his desk, Hong presents me with a postcard of the Renaissance Center in downtown Detroit—a futuristic-looking skyscraper with seven glassy, cylindrical towers in a ring. "A souvenir to help you remember bacteriorhodopsin!" When I tell him I don't understand, he smiles and shows me a computer-generated picture of BR. Seven helical columns that look like baloney curls stand in a ring around a light-sensitive pigment called retinaldehyde, or retinal A. "Retinal A is a close relative of the compound in our eye that helps us see in dim light. Nature is fond of reusing her winning designs in new ways," he says, as he pours me more green tea. "In BR, she uses an eye pigment to pull down the sun."

Staring at the skeleton sketch of BR, I imagine myself inside the Lilliputian columns when the sun breaks through the San Francisco fog. A photon zigzagging through the salty bay dives into the sensitive retinal A, causing it to shift shape, from straight to bent. As it kinks, the protein columns attached to retinal A are rattled as well. The amino-acid molecules studded throughout the rattled columns bump against one another, like passengers colliding in a lurching bus. The new proximity starts a handoff of a proton from amino acid to amino acid. In a nanoheartbeat, the positive charge moves from the inside of the membrane to the outside. A sunny morning can keep the handoff of protons working continuously.

What interests computer engineers is only the first part of this scene—the photon of light hits, and the molecule shifts. This flip-flop, from one state to another and back again, is automatic, even if the protein molecule is separated from its live host. "What most people don't realize is that you can remove BR from *Halobacterium* and embed it in plastic and it will work quite beautifully," says Hong. "In Russia, scientists have made a film of BR that they can still flip back and forth after fifteen years. That stability, we thought, would make it a good medium for storing information in computers."

Another of BR's talents is its knee-jerk reaction to certain frequencies of light—this means you can use one color of light to kink it (recording a one) and another color of light to unkink it (recording a zero). Here's how it works: In its relaxed state, BR will absorb only green light. If you flood it with green light, it kinks to a red-absorbing state. Then, if it's zapped by red, it unkinks, returning to the green-absorbing state. It's an endless toggle controlled by light.

This mechanism reminded computer scientists of the system already in use for storing digitized information. The working surface of magnetic hard drives or floppy disks is covered with tiny iron-oxide crystals, and they are able to flip their poles like little magnets. As the read/write sensors make their passes over different parts of the disk, they turn electrical signals into magnetic energy and vice versa.

In the case of optical protein computing, the working surface of the disk would be covered with BR molecules (*much* smaller than iron-oxide crystals) packed shoulder to shoulder. The read/write heads would be red and green laser beams, which, when aimed at specific "addresses" on the drive, would kink and unkink molecules, storing ones and zeros and then reading them out. An optical detector would measure whether or not light has been absorbed at each site. To keep from erasing information during the reading process, a second pulse of light would follow the read light to reset the flipped BR.

The thought of using a protein this small to store information quickens the pulses of computer engineers. Robert R. Birge, director of the W. M. Keck Center for Molecular Electronics at Syracuse University, went beyond dreaming and teamed up with physicist Rick Lawrence of the Hughes Aircraft Corporation in Los Angeles to flight-test a BR storage device. They laminated a thousand layers of BR, each a molecule thick, onto a thumbnail-sized quartz plate. "It looked like a piece of glass with a clear, deep, rich red coating," Birge said.

A laser was used to address not one molecule at a time (laser beams are still far too wide to do this) but a patch of about ten thousand molecules, flipping them all at once. Even in this configuration, says Birge, the device has a potential storage density of nearly ten megabytes per square centimeter, comparable to the storage density of elite magnetic devices available only in multimillion-dollar supercomputers. But that's only a beginning. When we find a way to focus the beams to write to each molecule, says Birge, a single 5¼-inch floppy disk coated with BR could theoretically hold 200 million megabytes (compared to the 1.2 megabytes that a disk that size holds now). Access times would be cut way down, too. It takes BR only five *trillionths* of a second to change absorption states. Give it a nanosecond, and it'll kink and unkink two thousand times, beating conventional magnetic devices by a factor of a thousand.

But in the speed-addicted world of computing, even this is not fast enough. Researchers at the Naval Research Center in Dahlgren, Virginia, are hoping to find or engineer a strain of *Halobacterium* with an even faster BR flip-flop. Ann Tate, manager of the Molecular Computing Group, explains, "When the BR molecule flexes from its unkinked to its kinked state, it goes through a continuum of shapes, each one with a different absorption spectrum. Right now, we concentrate on the ground state and the kinked state, and it takes five picoseconds to get from one to the other. What if we could

speed up the flexing? Find a BR that kinks at three picoseconds instead of five?"

The scientists hope to find that speedier BR in one of the millions of *Halobacterium* offspring they are raising in laboratory tanks.
Once they locate the winning microbe, they'll want to put its BR in
a storage medium that breaks all the records in terms of capacity.
That means going beyond two-dimensional BR films. "What we are
starting to do now is suspend BR in a Jell-O–like plastic that hardens
into cubes. When we get BR memory in 3-D like this, storage capacity will really balloon."

Imprisoning BR in a cube presented an opportunity but also a
logistical problem—how to read the molecules in the center of the
cube without the light beam triggering or destroying information on
the way in. Once again, BR's special qualities allowed engineers to
jog around this problem. Researcher Dave Cullin was explaining this
to me (with copious drawings) in the windowless belly of a Quonset
hut at naval headquarters in Dahlgren. "BR actually uses two photons when it photosynthesizes—it adds up the energy in the two.
This ability to absorb and combine two photons gave us an idea. We
could penetrate the cube with two rays, each entering from a different face, each of a frequency that, by itself, didn't affect BR molecules on the way in. At the point where the rays converged,
however, their frequencies would combine, and this energy would
be enough to write or read the data at that particular address." Dave
paused after this punch line, giving me time to admire the simple
ingeniousness of the two-photon scheme. For the thousandth time I
noticed my own tendency (a human tendency, I think) to be absolutely delighted by this sort of elegance. The same elegance that
nature, of course, has been choosing for eons.

So now that we have trillions of BR molecules in a device the
size of a sugar cube, what can we store? We could use BR just to
store zeros and ones, of course, but Robert Birge has a more ambi-
tious plan. He and his company, Biological Components Corporation, want to use the 3-D memory device to store analogue
holographic images in the BR, imprinting patterns of light and dark
instead of strings of zeros and ones.

A hologram is created by superimposing two beams of light onto
a piece of film. One beam of light contains the image, and the other
is plain light, called a reference beam. Where the light waves interfere on the film, they create a unique signature. The deconstructive
interference (where there is no image) causes dark areas and the

constructive interference (the image) is registered as areas of light. When you want to recall the original image, you simply flood the hologram with plain light—the reference beam—and it regenerates the original recorded pattern. In Birge's device, the film would be BR, and the light and dark patterns of the light waves would be recorded in kinked or unkinked molecules.

Holographic memory is especially suited to what's called correlation, or matching of images. You can take a picture of an airplane wing, for instance, and then fly the plane and take another picture. Comparing the two holograms would instantly show you where stress or strain has occurred. To make the holograms even more versatile, you can pass the images through a Fourier lens as you record them, which basically turns the image into a "frequency picture" so that the holographic correlator can recognize and match an object even if it is tilted at a different angle from the way it was when originally recorded. For instance, a pen would be recognizable whether it is held horizontally, vertically, or anywhere in between. (Our eyes have even more flexibility. We can recognize someone if they are close or far, or if their image is tilted side to side, forward or back. "Nature is ahead of us here," Tate admits, "but it gives us something to strive for.")

Fourier transforms made with conventional film can be layered like transparencies and held up to the light—when light shines through two of the transforms in the exact same spot, you have your match. What holographic BR memory can do, with mirrors and lenses, is place *hundreds* of BR-embodied Fouriers on top of one another to simultaneously find a match. This puts it streets ahead of digital techniques.

You could store pictures of all the customers in your bank, for instance, and when someone walked up to a teller, a camera would see the face and quickly match it to the hologram database, bringing up the customer's file. Even if the camera caught only an eye or the corner of a smile, it could recall the whole thing, because a hologram stores the whole in each and every part. If you wanted to do the same thing with a conventional silicon-matching device, you would first have to digitize the person's image into zeros and ones and then comb pixel by pixel for a string of numbers in your database that matched that person's numbers. In the holographic correlator, numbers are eliminated. You essentially put the entire stack of customer pictures on top of one another and look for the spot of light that shines through—signifying a match. This simultaneous search can be done so quickly that you could use a TV camera as an input device and identify people as they stroll through the lobby.

Information storage isn't a problem, either. If you figuratively sliced the cube into "sheets," you could store up to four hundred images per sheet, and then "pull up" a whole sheet at a time by slicing plain light through the cube to illuminate a cross-sectional slice. Even more images could be stored per page with a technique called angular multiplexing. By changing the angle at which the reference beam hits the cube, you could burn hundreds of holograms on the exact same spot and read them back with a tiltable laser.

If the system proves practical, Birge believes holographic memory could play an important role in robot vision, artificial intelligence, optical correlators, and other areas starved for complex pattern-processing capabilities. "This is an area where we could completely blow away semiconductors," he says. "We're going to be able to have the equivalent of twenty million characters of associative memory on a single film. You simply couldn't build a semiconductor associative memory with that many connections." And yet, I think to myself, an associative memory with many, many more connections has already been designed, and it's balanced on the stalk of my neck at this very minute.

After Conrad's compelling visions of self-assembling shapes bouncing in a maelstrom of motion, Birge's BR, as fantastic as it is, feels a little too confined—too on-and-off digital. To get back into more open spaces, I book a flight to the University of Arizona, Tucson, where I'm told I'll meet another biomimic who's determined to climb his own peak in the range of computing possibilities. In their ascents toward natural computing, Stuart Hameroff and Michael Conrad could easily run into each other on the trail.

According to Hameroff, the ultimate computer is not chemicals dancing in neurons, or light kinking proteins in a membrane, but rather the net of the spidery strands (cytoskeleton) assembling and disassembling in your cells as you read this. My survey of nature-based biological computing would not be complete without a visit to the man who sees the roots of consciousness in a microtubule. Buckle your quantum belts for this one.

THE SCAFFOLDS OF CONSCIOUSNESS?

Stuart Hameroff and I are in a stale canteen just off the operating room, waiting for him to be summoned, as he puts it, "to pass gas." At this moment, he looks more like the sax player on the cover of last month's *Downbeat* than an anesthesiologist. He's tilted chairback against the wall, feet up on the table, his scrub-green shower cap pulled low over a bushy ridge of salt-and-pepper eyebrows. At the back of his cap a ponytail struggles to break loose. He's staring at a green wall and talking a blue streak.

As my tape spools, his thoughts bank like swallows over a wide landscape: quantum physics, philosophy, computer science, mathematics, neurobiology (another person who needs a Dewey decimal system for his personal library). But he keeps circling back to the same subject, one that has tangled many a fine mind over centuries: the brain/mind debate. That is, does the mind float above and separate from the brain, or does it sprout from the gray goo itself? If it sprouts, by what biological mechanism does it emerge? And then, most mysteriously, how do these biological interactions inside the brain *converge* to afford us a "unified sense of self"—a single identifiable I?

In a few months, Hameroff will host an international think tank at the University of Arizona in Tucson on consciousness, a conference that already has several hundred registrants pawing at the chance to reenact the old debate. Hameroff has stepped into the consciousness fray in a public way lately, appearing in glossy magazine spreads with Roger Penrose, a mathematical prodigy known for his theories about wormholes, black holes, and geometric tiling. With his latest book, *Shadows of the Mind*, Penrose appears to have headed down a new wormhole, into the quantum world of biology-based consciousness. For a journey like this, Penrose decided, it's good to have a doctor along.

"I take away and revive people's consciousness every day," says Hameroff. "So I've thought about this in a very practical, nonabstract way. A biological way. We know, for instance, that certain structures in the brain physically change in the presence of anesthesia. That is, they stop moving when consciousness slips away. Wouldn't it follow that those same structures, and their movements, are tied to consciousness? Maybe they're the root of consciousness. I say they are."

The physical structures that Hameroff refers to are protein polymer tubes called microtubules, and amazingly, though they are thoroughly ubiquitous structures, appearing in every cell of our body, they were incognito until 1970. It seems we had been inadvertently dissolving them with a fixative (osmium tetroxide) used to prepare specimens for the electron microscope. (Don't you wonder what else we may be dissolving?) Once we realized how to prepare cells without destroying microtubules, we began to see them everywhere we looked, and it dawned on us how important they are.

Cells are not the droopy "bags of watery enzymes" that scientists once imagined. They are given their shape by the cytoskeleton—a Tinkertoy scaffolding of protein tubes and connectors that organize the interiors of all living cells. The protein tubes in this cytoskeleton are called microtubules, cylindrical fibers that can be anywhere from tens of nanometers long during early assembly to meters long in the nerve axons of large animals.

Microtubules are one of those examples of nature's geometric mantra repeated over and over. The building blocks of the microtubule are proteins called tubulin. Two varieties of tubulin, alpha and beta, self-assemble into dimers, which self-assemble end to end into long protein chains. These strands always group together in bundles of thirteen, forming a hollow cylinder made of protein. The cylinder's strands are twisted clockwise like twine in a rope, so that when the microtubules are viewed in cross section, they look like a child's pinwheel.

Each cylinder sports protrusions along its length called microtubule associated proteins, or MAPS. Some MAPS are bridges connecting the tubules to one another, forming the 3-D lattice that gives the cell its shape. Other MAPS, such as dynein and kinesin, are sidearm proteins (contractile spurs) that can extend and contract. Moving like the legs of a centipede, they act in a coordinated way to pass cytoplasm (cell fluid) along the tubule in bucket-brigade style, or to move organelles from one part of the cell to another. The cell's workers—chromosomes, nuclei, mitochondria, neurotransmitter synaptic vesicles, liposomes, phagosomes, granules, ribosomes, and the like—all ride the microtubule conveyor belt, meaning that microtubules are in on every just about every important cellular function you can think of.

Including reproduction. Remember those spindles forming and disappearing inside dividing cells in high school biology filmstrips? (I'm dating myself.) Those were microtubules helping to pull apart the doubling sets of chromosomes so that one cell could become two. Microtubules are also at work in cilia, the ubiquitous hairlike filaments that bacteria used to row themselves around your microscope slide. Cilia also line our mucous passages, and with the help of microtubules, they push materials up and down our body's smallest corridors. It's not an exaggeration to say that without microtubules, we wouldn't be able to sense the world, swallow, grow, or, says Hameroff, remember our names.

That's because brain cells are also full of these microtubular nets. Here they are not only conveyor belt and scaffolding, but also the builders and regulators of synaptic connections called dendritic spines. (The same spines that Donald O. Hebb said are responsible for opening a "dialogue" between two neurons so that learning can occur.) Microtubule assemblies are also present along the entire length of the spindly axon, and their branches are plugged directly into the neuron's all-important membrane and into organelles such as the soccer-ball–shaped clathrins at the end of the axon. These clathrins control the release of neurotransmitter chemicals, which swim across the synapse, delivering the neuron's signals. (In this last function, the microtubule has its finger in the very important pie of thought and feeling.)

Talking to Hameroff about the cytoskeleton makes you want to run into the streets and hand out pamphlets about this marvelous bio-invention. Here is a structure that should be a household word. It is a network nested within each neuron, which is itself nested within a larger neuronal net. The fractal beauty of this forest within a tree within a forest wasn't lost on Hameroff, and he began to wonder if there wasn't more to it. Perhaps the cytoskeletal net and the neuronal net are partners in the mind puzzle, working at different scales. Perhaps the tinier cytoskeletal net is the "secret basement" in the cognitive hierarchy, the root cellar of consciousness.

As Hameroff was finishing Hahnemann Medical School in Philadelphia and trying to decide what to specialize in, a professor told him that one of the effects of anesthesia was to cripple the microtubules in neurons. He now says, "That made me think. Is there a mechanism in microtubules that controls self-awareness, intuitive thought, emotion? Do microtubules help power consciousness?" Hameroff specialized in anesthesiology and began to read everything he could about the gas's chilling effect on microtubules.

Another revelation came years later when colleague Rich Watt brought him an electron-microscope portrait of a tiny network and said, "What does this look like?" "Cytoskeleton," Hameroff shot back, which made Watt smile. "Look again," he said. "It's a microprocessor, a computer chip."

The eerie resemblance had a profound impact on Hameroff. "The structure of the cytoskeleton is not coincidental, I decided. And the fact that consciousness fades when the microtubule quiets is not coincidental. The cytoskeleton network is as parallel and as interconnected as the neuronal net, but a thousand times smaller. It contains millions to billions of cytoskeletal subunits per nerve cell! The cytoskeleton, I decided, is a lot more than mere cell scaffolding or a protoplasm traffic cop—it's a full-fledged signaling network—a processor for coding, storing, and recalling our flickering thoughts. In short, it's biology's computer."

Quantum Leaps

For ten years, when he hasn't been escorting people in and out of consciousness, Hameroff has been modeling tubulin arrays on his computer, searching for some sort of code and signaling mechanism. "Do you have a minute?" He crooks a finger and then he's careening down the hallway, like a New Yorker on his lunch hour, to the media instruction lab where he's asked a biological illustrator to create an animated cartoon of flexing microtubules for the upcoming consciousness conference.

As it plays, Hameroff narrates, excited to see the world that has lived for so long in his imagination perform in living color, even if it is only a cartoon. "Each microtubule is a hollow cylindrical tube with an outside diameter of about twenty-five nanometers and an inside diameter of fourteen nanometers. Each tubulin dimer is about eight nanometers by four nanometers by four nanometers, and consists of two parts, alpha-tubulin and beta-tubulin, each made up of about four hundred fifty amino acids."

In the cartoon, a single dimer is pulled out and magnified—it looks like a C character in the fat outline font. "At the elbow of the C, the junction of the alpha-tubulin and the beta-tubulin, there is a hydrophobic [water-fearing] pocket. In this pocket, an electron moves up and down in a metronome ticktock fashion called dipole oscillation. As it oscillates, it changes the shape of the protein, crimping the C and then stretching it."

As we watch, cartoon beads of anesthesia gas start infiltrating from screen left. "Count backward from one hundred," mumbles

Hameroff. The minute the gas beads reach the dancing dimer bean, the electron in the pocket freezes, and the dancing stops. "Goodbye consciousness," he announces.

Based on his own observations, Hameroff now believes that the electron freeze is caused by anesthetic molecules jamming into the hydrophobic space at the elbow of the C and binding there. When the electron stops oscillating, we lose consciousness.

But it's not just the consciousness of higher animals that is affected by gas. Anesthesia can also stop the movement of paramecia, amoebae, and green slime molds, all of which rely on cytoskeleton for their oozing-forward movement. Hameroff knew that electrons acting alone inside each tubulin couldn't possibly account for something as coordinated as a paramecium moving to catch its prey, let alone a conscious thought. Somehow, he theorized, the oscillating electrons must *cooperate* in a larger signaling and communication network. To find a plausible mechanism, Hameroff looked to a theory of computation known as cellular automaton theory.

A cellular automaton computer is a software program that sets up a grid of squares or "cells" (the spreadsheet kind, not the living kind). Each cell has a definite number of neighbors and has a formula of sorts embedded in it. The formula is called a transition rule. At discrete time intervals, a kind of musical chairs occurs. Every cell must check out the status of all of its neighbors and then change states—either on or off—according to its transition rule. A rule may state: If at least four of my six neighbors are "on," I'll be on too. Otherwise I'll stay off. At each tick of the computer's clock, the cells check out their neighbors and change on or off accordingly. It helps to think of the "on" squares as white and the "off" squares as black.

Amazingly, simple rules and a clock regulating the action lead to regular patterns of white and black developing and moving across the grid, in the same way that "The Wave" can propagate through a crowded stadium of strangers. With more complex rules, a cellular automaton in three dimensions could simulate the formation of a snowflake, mollusk shell, or galaxy. In fact, John von Neumann, known as the father of modern computing, suggested in the 1950s that such a lattice could be programmed to solve any problem. Learning this, Hameroff wondered, could microtubules be doing something like The Wave on their latticework of tubulin? Could they somehow be computing?

The illustrator fast-forwards the computer animation to a functioning array of microtubules. For this demonstration, he's slit the soda straw of a microtubule lengthwise and unfurled it flat into a rectangular array. Each C-shaped tubulin is resting in spoon formation with its neighbors, so that the state of each dimer (whether its electron is up or down in the pocket) could be affected by the electrostatic state of its six neighbors. He hits PLAY, and an excited vibration begins in a patch of the array at one corner and ripples across the array like the energy of a wave moving through water. But it doesn't stop there.

Hameroff believes that a microtubule can "catch" the oscillation of its neighbors—that is, a set of proteins vibrating in one microtubule could start another set vibrating in exactly the same way, like a tuning fork starting to vibrate in response to another in the same room. This "catching" oscillation, says Hameroff, may be possible because of a very unusual set of qualities that make microtubules the perfect substrate for quantum coherence.

"Coherence" is a hyper-organizing that imparts a strange and often wonderful quality to ordinary matter. When the crystals in a laser rod are pumped with enough energy, for instance, they will all of a sudden vibrate in lockstep fashion, and give off coherent laser light. Or when the linked electrons in a metal take on identical quantum characteristics, they become nearly frictionless conductors (a phenomenon called superconductivity). In supermagnets, microdipoles align, and in superfluids like helium, quantum-synchronized atoms create a friction-free fluid. But superconductors, supermagnets, and superfluids typically require temperatures near absolute zero to dampen thermal noise and bring their particles into alignment. The question is, can coherence happen in biological materials, at bodylike temperatures?

In the 1970s, Herbert Fröhlich of the University of Liverpool postulated that electrons trapped in the hydrophobic pocket of a protein like tubulin could oscillate, causing the protein to change shape in a predictable way. Further, he predicted that these electrons would oscillate *coherently* if they were in a uniform electromagnetic field (such as the walls of a microtubule) and were pumped with enough energy (provided by the bond severing of molecules like ATP or GTP). At some point, a set of proteins could reach a critical level of excitation and all of a sudden align in lockstep.

Applied to the microtubule, Hameroff postulated that the pattern of oscillation could either travel in waves, rippling across the lattice, or jump to nearby microtubules. These traveling shape-changes could allow signals to be carried throughout the neuron—

signals that could direct, for instance, the movement of cilia or even the regulation of synaptic strengths. But how far could this coherence reach? If it could spread microtubule to microtubule, could it also go outside the neuron's walls?

Consciousness, a brain-wide phenomenon, cannot be isolated to a neuron or two. In order to explain the "unified sense of self," microtubules would need some way of coordinating their actions across large distances in the brain. To explain unity of self, Hameroff wandered even farther into the labyrinth of quantum mechanics.

As part of his exploration, he read a book by Roger Penrose called *The Emperor's New Mind*, in which Penrose found quantum theory a thoroughly plausible explanation for how thoughts can appear to be magically distributed or "floating above" the brain, and yet still be anchored in matter. According to Penrose, if we could find the biological player in this quantum dance, we might be able to explain the unified sense of self.

Quantum mechanics applies to the very small things in our world, the substructure that underlies the visible world. In the early decades of the century, when quantum mechanics was first taking shape as a theory, it completely upended our ideas of physical reality. Newtonian laws were not completely banished—they still applied in our *visible* world—but they were no longer the be all and end all. Newton had no idea how weird the world of the tiny could be.

Two relevant legs of the quantum theory are the "superposition of states" and "quantum knowing." The theory of superposition says that atoms are in many possible states simultaneously. They are searching among the various alternative energy states (an effect Michael Conrad called "quantum scanning"), and they don't "choose" a state until they collide with matter or are observed. The famous argument in support of this is provided by the double-slit experiment, in which a low-intensity beam of photons is projected onto a wall punctured with two vertical slits. Behind the wall is a screen. Because the intensity is low and the photon stream is "dilute," each photon should pass through one slit or the other. Instead, the pattern on the screen suggests that each photon passes through both slits at once. The bizarre but oft-replicated experiment seems to suggest that a photon can be in two places simultaneously.

Quantum theory says the photon is not just in those two places, but in many others as well. Scientists decided the best way to talk about a photon's location would be to imagine a three-dimensional graph of all possible states. This is called the state space, and the "wave function" is a way of characterizing all the possible states that the photon may be in. Amazingly, when a particle comes into contact with matter—the molecules on the screen in the famous two-slit experiment, for instance—the wave function "collapses" to a single point, and the photon is forced to choose a single state to be in. When we observe something, we don't see all its possible states—we see only one. We force it to be in only one state through the act of seeing or measuring it.

Michael Conrad has suggested that biological molecules exploit this freedom to shuffle the deck of many possibilities and explore possible solutions to, for instance, the problem of shape-based docking. In his view, enzymes are physically flopping around just before docking, and an electron tries out many different bonds, searching for a minimum energy configuration. Penrose postulated that our creative minds may play with possibility space in the same way—trying out dozens of different options simultaneously until one emerges as a conscious thought—a decision about what state to be in.

The second quantum theory that seems to relate to "mind" is the idea of quantum knowing. This states that movements of atoms, electrons, or other quantum particles may, under certain instances, be synchronized at great distances. As Hameroff writes, "The greatest surprise to emerge from quantum theory is quantum inseparability or nonlocality which implies that all objects that have once interacted are in some sense still connected! Erwin Schrödinger, one of the inventors of quantum mechanics, observed in 1935 that when two quantum systems interact, their wave functions become 'phase entangled.' Consequently, when one system's wave function is collapsed, the other system's wave function, no matter how far away, instantly collapses too."

Talk about a truly interconnected world! Naturally, quantum knowing has been applied to many theories of cognition, including the holographic model of consciousness. Quantum knowing says that once two particles have been entangled quantumly, been part of the same quantum wave function, they are always related in some way—they know what their coherent relative is doing. In a sense they *are* their correlate particle. This means the same coherence that causes patterns to oscillate in synchrony inside the microtubule may cause coherence to occur in quantum relatives clear across the brain (or across brains!), without the need for neurons to be touching. Perhaps this same quantum knowing may account for such "supernatural phenomena" as Jungian collective unconscious, Hegel's world spirit, and the strange ESP that you feel with a loved one who is miles away.

At the time he wrote *Emperor*, Penrose had the quantum arguments for consciousness worked out, but knew of no biological mechanism in the brain that would be capable of such quantum effects. He speculated that quantum effects in the brain would require a structure that was 1) small enough to be driven by quantum effects and 2) separated from the thermal hubbub of the rest of the brain. When Hameroff read these words, he found himself talking back to the pages. Tubulin proteins *were* small enough to host the quantum effects Penrose so beautifully described, and the hydrophobic cages inside the fibrils would indeed be a safe haven from the rest of the brain! He was ecstatic. "Penrose had handed me the quantum argument that I had been searching for, and I believed I was holding the missing biological piece that he needed."

Hameroff wrote to Penrose and asked to come and see him. At a famous two-hour mindmeld at Penrose's Oxford office, the two exchanged the missing pieces of the conceptual locket each had been carrying around. A few weeks later, Penrose stood up at a meeting and postulated that the microtubule may be the physical seat of consciousness.

In his latest book, *Shadows of the Mind*, Penrose lays out his arguments in a formal way. He believes that "mind" is a "macroscopically coherent quantum wave function" in the brain that is protected from entanglement with the thermal environment. The wave function is composed of quantum-connected electrons sitting in superposition—at both the upper and lower position of the hydrophobic pocket of each protein dimer. Because the pulse of vibrational energy in a microtubule is separated from the hubbub of the brain, it isn't forced to choose a single state, and is free to investigate all possible patterns.

Penrose and Hameroff believe the microtubules' almost crystalline structure may allow them to support a superposition of coherent quantum states for as long as it takes to do "quantum computing." When the quantum superposition finally collapses, it triggers a spontaneous release of neurotransmitters (microtubules also direct this process). With this release, a thought, image, or feeling occurs to us. At this point they are trying to figure out how many neurons it would take for a conscious event—a collapse—to occur. They think the number may be ten thousand cooperating neurons.

As if coherence and cellular automata are not enough, Hameroff has entertained some half dozen other theories about how signals may be bounced around the brain on the trampoline of tubulin. Another theory imagines that the hollow tubes act as waveguides, like little fiber-optic cables. The water within the tubes structures in such a way as to emit a photon, which bounces along the waveguides, creating a tiny optical computer within our cells. Cytoskeletons may also be using soliton waves, sliding motions, coupling of calcium concentrations to cytoplasmic sol-gel states, or constant polymerizing and depolymerizing to process signals.

Regardless of how microtubules are computing and communicating, Hameroff is convinced that they are, and he thinks if we let microtubules assemble themselves in a laboratory, we could get them to compute for us. "The neat thing about microtubules," he tells me, "is that they can function outside of their cellular home [like BR can]. Put tubulin subunits in the right solution and they do what comes naturally—they self-assemble into beautiful cylinders crosslinked with MAPS. That means we could conceivably grow arrays of them in vats and use them as signaling media. We could use them as a storage device or even as an intelligent processor."

Michael Conrad is also interested in this cellular trellis that so recently showed its face under our microscopes. "Chances are that microtubules will be a part of the tactilizing processor someday," says Conrad. "Using their tiny centipedelike arms, they could push or pull the shapes, speeding them toward one another for self-assembly into a mosaic. Cytoskeletons could even be part of the readout mechanism. Instead of forming a mosaic, the floating shapes could somehow influence the self-assembly of cytoskeleton. The final shape of the cytoskeleton would reflect the pattern of inputs to the neuron [it would say "snowshoe hare"], and the readout enzymes would interpret the cytoskeleton instead of the mosaic. Finally, you could hook microtubules into long strands that act as physical transmission lines—wires—to connect tactilizing processors to one another in complex parallel networks."

For Conrad, the cytoskeleton is like having a new multitalented personality join the team. "Think of all the processes that cytoskeletons may employ—conformational change, dipole oscillations, sliding motions, soliton waves, vibratory motions, sound waves, polymerization and depolymerization! This gives the system a lot of dynamics to work with, a lot to choose from when evolving a more efficient way to compute. Our idea is to feed evolution all the flexibility we can, and then stand out of its way and let it seek its own opportunities."

Stuart Hameroff has written a book-length ode to microtubules called *Ultimate Computing*. He is bold in print, and his monograph is a fascinating romp. It plunges into difficult mathematics and then suddenly breaches into the stratosphere, making predictions that raise eyebrows in some scientific circles. A cytoskeletal array would be a fine medium for artificial intelligence, Hameroff contends. How quickly could it compute? Well, in the three-and-a-half-pound universe we call a brain, there are 10^{15} tubulin dimers, each operating at a speed of about 10^9 operations per second, for a total of 10^{24} operations per second. If you want more dimers than that, make a bigger vat! We could even, he says with a bravado that has earned him some brow bends from more cautious colleagues, send vats of this stuff into orbit around the Earth, where it can evolve artificial consciousness. Or, he says, because microtubules are biological molecules, they would be welcome in our bodies. We could capitalize on their motorlike MAPS, he says, and send them as programmed nanorobots to do specific tasks inside the cell.

Hameroff ends his book with a dare: "Microtubules and the cytoskeleton created their place in evolutionary history by being problem solvers, organelle movers, cellular organizers, and intelligence circuits. Where do they go from here?" In a *Computer* magazine article in 1992, Hameroff and four other authors wager a guess: "If computation occurs in microtubules and can be decoded and assessed, cytoskeletal arrays may provide 'devices' with substantial computing power. Perhaps such systems will someday reach cognitive capabilities comparable to and even superior to human abilities." And then the authors seem to read our minds: "While the ideas of dynamic coding and technological intervention in the cytoskeleton may seem farfetched, are they any more radical than were the ideas of static genetic coding and intervention in DNA and RNA some years ago?"

I thought of Hameroff's claim when I read an article called "On the Path to Computing with DNA" by David Gifford in the November 11, 1994, issue of the journal *Science*. Someone had to think of it sometime. If simple enzymes can compute through shape-fitting, as Conrad contends, and if Hameroff's microtubules can assemble and disassemble to form computing arrays, then what about the most wondrous coding mechanism of all: the code of life that twines together like two circular staircases, pairing up base by base in a simple yet splendid feat of pattern recognition? It was only a matter of time before someone climbed the peak to the DNA computer.

TRAVELING SALESMAN, CONSULT YOUR DNA

DNA is a code, a kind of language, and you can translate what you want to say into the four-letter alphabet of nucleotide bases: A (adenine), T (thymine), G (guanine), and C (cytosine). By turning your information into a chain of molecules, you've managed to turn it into something that can be touched, something that's controlled by the physics of shape-fitting and sequence matching.

You've also turned it into something that can be duplicated, in part because of a neat rule about complementary DNA. Here's how complementarity works: When two strands of DNA get together, their bases line up very specifically. An A sticks to T, a C to G, and so on. Since combining is the energetically favorable thing for complementary strands to do, they will always zip together into the double-spiral helix of Crick and Watson fame. You can heat them to split them apart, but let the solution cool back to body temperature and they'll rejoin without skipping a beat. Kevin Ulmer of Genex Corporation in Rockville, Maryland (now of seQ, Ltd.), says that's like taking a Chevy apart, sticking the parts in a large crate, shaking it up, and having it reassemble into a car you can drive away. Since DNA "processors" are a little smaller than Chevys, however, they can fraternize by the trillions in a thimbleful of water, making them ideal for parallel processing.

DNA's propensity for automatic assembly gave Leonard M. Adleman, who holds the Henry Salvatori Chair in Computer Science at the University of Southern California School of Engineering, an idea. In 1994, with a few test tubes of synthesized DNA strands, he set out to solve one of the most difficult computing problems known. The "directed Hamiltonian path problem" (finding a sleek path through a network of points) is a benchmark for computer prowess because an efficient algorithm (means to a solution) has yet to be found. The problem is that of the traveling salesman who must fly to many cities, but who wants an itinerary that will take him through each city only once. When there are many cities, the possible itineraries become astronomical. A trillion-operations-per-second computer trying to find a Hamiltonian path through one hundred cities, for instance, would need 10^{135} seconds—vastly longer than the age of the universe!

Adleman used only seven cities, looking for a path that would begin in Atlanta, end in Detroit, and pass through each intervening city only once. He gave each city a DNA name, using the letters of the DNA alphabet, A, T, G, and C, and then set out to create strands of DNA that would complement these names. To create these strands, Adleman used an increasingly common piece of lab equipment called an oligio machine that strings bases together automatically. As you'll see in the third column below, he replaced each A with a T, each T with an A, each C with a G, and each G with a C, according to the rules of complementarity.

CITY	DNA NAME	SYNTHESIZED COMPLEMENTARY DNA NAME
Atlanta	atgcga	tacgct
Baltimore	cgatcc	gctagg
Chicago	gcttag	cgaatc
Detroit	gtccgg	caggcc

(Adleman actually used seven names with twenty letters each, but we'll keep it simple.)

Using common recombinant DNA technology, Adleman made 30 trillion copies of these complementary DNA strands and set them aside.

Adleman then gave each segment of the route a flight name—taking the last three letters of the departure city and attaching it to the first three letters of the arrival city. If he were using English, the Atlanta to Chicago flight name would be the six capital letters in the following example: atlaNTACHIcago. But since Adleman was using DNA code, the flight names looked like this:

FLIGHT	DNA NAMES	DNA FLIGHT NAMES
Atlanta–Chicago	atgcga–gcttag	**cgagct**
Chicago–Detroit	gcttag–gtccgg	**taggtc**
Chicago–Baltimore	gcttag–cgatcc	**tagcga**
Baltimore–Detroit	cgatcc–gtccgg	**tccgtc**

Using the oligio, he manufactured the DNA flight names in actual bases, then made 30 trillion copies of each. The idea was that if stirred into the same test tube, these flight names would stick to the ending of one city name and the beginning of another, thus splinting the two names together. To test this, Adleman poured the flight names into the test tube of complementary DNA city names. (So far, lab technicians assure me, it's as easy as Hamburger Helper.) Sure enough, the flight strands acted as splints; cgagct floated over to Atlanta and Chicago, for instance, and stuck to them like so:

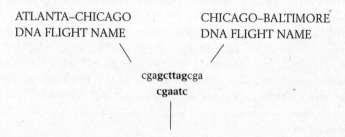

ATLANTA–CHICAGO   CHICAGO–BALTIMORE
DNA FLIGHT NAME   DNA FLIGHT NAME

cgagcttagcga
cgaatc

CHICAGO COMPLEMENTARY DNA NAME acting as a splint

It didn't take long before the test tubes were full of long strands of DNA flight names splinted together. Through a series of recombinings and screenings, Adleman was eventually able to filter out all the strings that started or ended with the wrong city, or were too long or short. He was left with only strings of DNA molecules that represented the winning itinerary.

The problem was solved through self-assembly, the kind that occurs in Conrad's shape-based computing. As David Gifford commented in *Science*: "The 'oracle' in Adleman's method is the immense computational capacity of a ligation reaction that produces billions of products and by brute force tries all possible solutions."

In Adleman's first experiment (detailed in the *Science* article), only seven cities were chosen, but it seems clear that almost any Hamiltonian path problem could be solved this way. But it's not just travelers' queries that stand to be answered. Complex problems such as telephone network switching, automating factory tasks, and artificial intelligence require what is called simultaneous processing. While conventional computers can explore only one or two solutions at a time, trillions of DNA molecules, each acting as a processor, can generate billions of possible solutions simultaneously.

In his press release, Adleman cheers cautiously: "It is premature to judge the long-term implications of this approach to computation; however, molecular computation has certain intriguing properties that warrant further investigation. For example, while current supercomputers can execute about a trillion operations per second, molecular computers conceivably could execute more than a thousand trillion operations per second." In fact, it has been estimated that a DNA computer could perform more operations in a few days than all the calculations ever made by all the computers ever built.

He goes on to write: "Further, molecular computers might be as much as a billion times more energy efficient than current electronic computers. Also, storing information in DNA requires about 1 trillionth the space required by existing storage media such as video tape. . . . For certain intrinsically complex problems . . . where existing electronic computers are very inefficient and where massively parallel searches can be organized to take advantage of the operations that molecular biology currently provides, it is conceivable that molecular computation might compete with electronic computation in the near term."

A few months after the paper was published in *Science*, Adleman held an impromptu conference on DNA-based computing in Princeton. To his amazement, two hundred scientists packed into a standing-room-only hall. Many talks were given and plans hatched, and though Adleman contends the field is still in "an embryonic stage," others at the conference believed we might see some practical DNA computers in as little as five years.

Chances are, silicon computers won't be abandoned completely—like Conrad's tactilizing processors, DNA enthusiasts see vats of DNA as souped-up peripherals for silicon computers. They'd make a tremendous storage medium, for instance. One speaker said that a liquid DNA computer one cubic meter in size could memorize more information than all the existing computers in the world. Getting down to specifics, Eric Baum of the NEC Research Institute at Princeton estimated that one thousand liters of DNA solution could contain in coded form 10^{20} (that is 1 followed by 20 zeros) "words" of information. Another speaker estimated that a million times more information could be stored at the bottom of a test tube of DNA than in the entire human brain.

Members of the press gasped audibly at these forecasts, given the fact that Adleman's experiment actually worked. It prompted Steven Levy of *Newsweek* to write, "Such an event is the equivalent of peering out the window of a bullet train and watching in astonishment as an unfamiliar vehicle zips by you with a fearsome whoosh. As if you were standing still."

But for Michael Conrad or Stuart Hameroff or Ann Tate, Adleman's announcement was an inevitable event, the shape of things to come. As Adleman has since remarked, his experiments have made him realize that "being a computer is something that we externally impose on an object." He suggests there may be a lot of other "computers," like DNA, that we have yet to discover.

Indeed, we are just beginning to investigate all the ways nature has already found to compute and transfer information. What may be most surprising is that it has taken us this long to look over nature's shoulder for computing ideas. Perhaps it's because our "search image" has been wrong; we haven't "seen" nature's computing devices because they don't look like ours.

Not yet, anyway.

TO UNFLATTEN BIOLOGY:
THE REAL QUEST

When I ask Michael Conrad what desktop computers will look like in the era of molecular computing, he hedges. For him, the real carrot is not the device. "The last thing the world needs is another new device," he says. "As an aesthetic thing I can understand technology, but except for some medical technologies, I don't really see technology as a human need. Our perceived need for technology is mostly generated by the competition of countries for export. I think it's economies, not people, that need devices in order to grow." This man, the head of a major computer center, doesn't drive a car, nor does he need to. He walks to work from the Victorian apartment he and his wife, Debby, have lived in for fifteen years. If he misses a phone call while he's walking, he doesn't know about it; he's beeper-free.

Conrad's agenda, and the prime directive in his vision for the future, astoundingly enough, is to offer people a new paradigm by which to understand biology—a biological rather than a mechanical paradigm. "Right now, this Mac Plus is the ultimate machine—it's what we know. That doesn't mean we should use it to explain the brain."

He's right. We have a habit of making theories about organisms and basing them on the machine of the hour. We used to say that the human body worked like a clock, but that was when the clock was the ultimate machine. There was also a time when we said it worked just like levers and pulleys and hydraulics. Then we said it was like a steam engine, with a distribution of energies. After the Second World War, when we began to devise feedback controls for our factories, we said our body worked like a self-regulating governor or servomechanism. Now, predictably, we're convinced that the body works like a computer. We're using theories from computer science—theories that come from the machine world—to explain how the brain works, and that disturbs Conrad.

"We are teaching biology students that our enzymes and neurons are simple switches, turning on or off. In reality, we're nothing like a computer, nor are we like a clock, a lever, a servomechanism, or a steam engine. We're much more subtle and complex than that.

"This view of the organism as a digital computer has flattened biology, and I'd like to unflatten it. When I build the tactilizing processor, I hope it will make people stop and consider that there is more than one way to compute. Nature's computers don't work the way ours do. To think that they do is very bad for society—it makes us use digital computers for tasks we ought to be asking our brains to do—tasks to which digital computers are not suited."

I thought a lot about what Conrad is trying to accomplish, and I think it's much more important than beating other countries to the sixth-generation computer. Conrad's insistence on unflattening biology reflects biomimicry's ultimate goal—to learn more respect for nature and to recapture our sense of wonder. At its best, biomimicry should take us aback, make us more humble, and put us in the learner's chair, seeking to discover and emulate instead of invent.

In their respective books *The Death of Nature* and *The Reenchantment of the World*, Carolyn Merchant and Morris Bergman agree that only by changing our perception of nature will we change how we behave toward her. There's a history that proves them right. In the 1700s we ignored cultural taboos about violating nature and gave scientists permission to break the natural world into pieces to study it. With the animus and mystery gone, nature was suddenly on our leash, to do with her as we pleased.

Two centuries later, having taken reductionism about as far as we can go, there are signs that a rebound is beginning. Many scientists, especially those in the ecological sciences, have become students of the whole once again. Attitudes toward nature have also come full circle, reanimating life and restoring reverence to our relationship with the natural world.

In concert with all this, the biomimics are showing us that nature is the ultimate inventor, and that there is much that we as observers do not know—perhaps cannot know. By forming alliances with her, by using biology-friendly materials and letting evolution work its magic (even without knowing how it works), we're bound to come out ahead of where we would be with our own linear, digital, rigidly controlled logic.

Will we be able to replicate exactly what happens in our brains by using carbon-based devices like the tactilizing processor, the microtubule array, a cube of BR, or a thimbleful of DNA? Michael Conrad laughs. "Remember, I have no illusions. I come from an origin of life lab and I know how fantastic life is. To emulate nature, our first challenge is to describe her in her terms. The day the metaphors start flowing the right way, I think the machine-based models will begin to lose their grip. Natural processes and designs will finally be the standard to which we aspire. On that day, I'll feel like I've done my job."

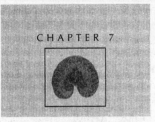

CHAPTER 7

HOW WILL WE CONDUCT BUSINESS?

CLOSING THE LOOPS IN COMMERCE: RUNNING A BUSINESS LIKE A REDWOOD FOREST

When we objectively view the recent past—and two hundred years is recent even in terms of human evolution and certainly in terms of biological evolution—one fact becomes clear: The Industrial Revolution as we now know it is not sustainable. We cannot keep using materials and resources the way we do now. But how are we to land softly?

—BRADEN R. ALLENBY, research vice president,
Technology and Environment, AT&T

Nature has evolved systems over billions of years that work in harmony with each other, that build from bare, rocky, thin soil to lush, green forests. Without human intervention the processes of nature have evolved self-regulating forces of beauty, grace, and efficiency. Our challenge is to learn how to honor them and be inspired by their truth to create new cultural values and systems.

—JAMES A. SWAN and ROBERTA SWAN, authors of
Bound to the Earth

Stewart Brand, editor of the first *Whole Earth Catalog*, calls himself a "lifelong purveyor of the biological metaphor." As a collector of tools and tips for the back-to-the-land-cum-sustainable-living movements, he realized long ago that the best tools are those that nature has already invented. So naturally, when Brand heard business consultant Hardin B. Tibbs talk about remaking industry in nature's image (at the 1992 EcoTech Conference in Monterey, California), he wanted to be part of it. He stepped up after Tibbs's talk and offered him a job with the Global Business Network, a consulting company working toward a sustainable economy.

Tibbs is one of the evangelists of the new movement called industrial ecology, my vote for biomimicry's most oxymoronic term. Coiners of the phrase hope that one day it won't be ironic, but will instead be an accurate description of how we conduct what author Paul Hawken calls the "ecology of commerce."

Considering its esoteric roots (my first contact with the term industrial ecology was in the *Whole Earth Review*, next to a testimonial on psychoactive plants), imagine my surprise when I read that AT&T, the fifth-largest company in the world, was sponsoring industrial ecology conferences, giving out industrial ecology fellowships, and creating a whole department in Bell Labs to wrestle with the concept. Then I read that General Motors was signing up, and that President Clinton's U.S. National Technology Strategy would feature industrial ecology as its guiding principle.

The mainstreaming of a very radical idea is in the works, and if it works, it promises to change far more than the way we make computer chips or fiber or adhesive. It promises to change the way we make, sell, market, and buy everything. As strange as it seems, industrial ecology will conduct business the way a sun-soaked hickory forest recycles its leaves.

FROOT LOOPS AND THE FUTURE OF THE HUMAN RACE

When I first saw Bob Laudise, adjunct chemical director at AT&T's prestigious Bell Laboratories, he was at a podium, waving a box of Froot Loops above his head so that neck craners eighty rows back could get a better look. The room held a thousand or so inventors, scientists, and manufacturers from some of the largest companies in the country—producers of electronics, high-tech materials, and durable goods. After a buildup worthy of Barnum, Laudise dropped down into the aisle and handed a few boxes around. Men and women in conference attire eagerly passed them to one another, lowering glasses down their noses to read the ingredients. They conferred in serious tones and jotted notes in leather portfolios: Re: Froot Loops—find secret ingredient.

The secret Laudise referred to was a new way to clean electronic circuit boards—those platforms of tiny transistors and other components that control electronic devices. Right now, hazardous toxic solvents are used to clean the boards between manufacturing steps. The AT&T researcher who devised the new cleaner was inspired by the basic tenet of industrial ecology, which says that we should try, wherever possible, to work only with substances that nature would recognize and be able to assimilate. Taking this idea quite literally, the researcher had parked himself in front of a database of FDA-approved substances and identified a slurry of ingredients so benign that kids could slurp them from the bottom of their cereal bowls. Yet when poured on a freshly manufactured circuit board, they washed away the leftover solder and other gunk like a lucky charm.

The question is, why haven't we always worked with something nature-compatible? Wouldn't that have avoided a lot of problems? Amazingly, it took a drastic realignment of our thinking to come around to adopting this simple tenet. One hundred years into the Industrial Revolution, we are only now opening our eyes and realizing that our artificially constructed world is not isolated from the real one. It is enmeshed in a larger natural world that cradles and nourishes us, making all of our activities possible. Fouling this nest, a lesson other organisms learned long ago, can be a deadly business.

PURSUING FOLLY TO THE LIMIT

At first it was hard to see that we were fouling our own nest—we kept expanding into fresh new territory and leaving our tired land and waters behind. It was as if we were a small seedling growing rootlet by rootlet into a fragrant pot of soil. All was fine as long as the rootball of our economy, our world within a world, was small in relation to the larger natural setting.

Unfortunately, we didn't stay small, and the natural world, of course, didn't get any bigger. It doesn't take a Malthusian to tell us that we have grown to fill our container. Each month, 8 million people (the population of New York City) join an Earth that is already groaning. In the United States alone, we generate 12 billion tons of solid waste a year—that's twenty times the total amount of ash released by the 1980 eruption of Mount St. Helens! Over 200 million tons of airborne wastes are added to the atmosphere each year, joining the 90,000 tons of known nuclear waste, most of which will be poisonous for another 100,000 years.

Our industrial resource cycles now rival or even exceed the Earth's biogeochemical cycles. As Tibbs reports, "The industrial flows of nitrogen and sulfur are equivalent to or greater than the natural flows, and for metals such as lead, cadmium, zinc, arsenic, mercury, nickel, and vanadium, the industrial flows are as much as twice the natural flows—and in the case of lead, eighteen times greater."

It's not just the magnitude of the numbers that frightens—it's the rate at which they are accelerating. Consider that it took from the beginning of human history to the year 1900 to build a world economy that produced $600 billion in output. Today, the world economy grows by this amount every two years.

So, if we were once a tiny seedling in a fragrant pot of soil, we are now horribly rootbound, pushing dangerously close to the edges of nature's tolerance. How is it that we did not see this coming?

Braden Allenby has wondered this himself, and in the introduction of his Environmental Sciences doctoral thesis, he describes quite beautifully how we crafted and mounted our own blinders. He spends the rest of the thesis showing us how to take them off, and how to change course with an approach that has roots in ecology. I went to visit Allenby at his Bell Labs office, where, as research vice president of technology and environment, he is paid to spin ideas like globes in his hands, looking at them every which way.

Allenby is dark-haired, bright, and intense; he speaks in a swift current, drawing patterns in the air, and sweeping you along like a storyteller. For millions of years, he tells me, there were simply not many of us, and our impact was limited. There were taboos against truly invasive practices. (As Carolyn Merchant notes in her book *The Death of Nature*, nature was seen as a living entity, a mother, and it was deemed unthinkable to cut mother's hair [deforestation] or penetrate mother's bowels [mining].) In the seventeenth century, says Allenby, mores began to change. The Scientific Revolution made reverence for the Earth obsolete, while the Church condemned it as druidic superstition. Once nature was demoted to a dead and soulless assembly of atoms, it became socially acceptable to exert our "God-given" dominion over her. The path was cleared for worldwide exploitation.

Still, insists Allenby, when biceps and back muscles ran the shovels, our rate of destruction more closely matched nature's rate of renewal. It wasn't until the Industrial Revolution put us on the winning side of a very large lever that we began vaulting past nature. Gears, hydraulics, fossil fuels, and the internal combustion engine allowed us to tap deeper, faster, and farther into the Earth. We began to extract resources as quickly as we could, transforming them into products, waste, and, of course, more people. The farther removed we became from nature in our attitudes, lifestyles, and spirituality, the more dependent we became on the products of this transformation. We became addicted to the spoils of our "rational mastery."

Still, physical limits seemed far away. We were in a colonizing mood, confident that vaster territory and richer riches lay just over the hill. With virgin materials nearly free for the taking, there was no point to recycling or reusing what we had extracted, nor was there any reward. In fact, the fledgling science of economics measured the well-being of a nation by its "throughput": how many resources it could transform each year, and how fast. In the nation-against-nation scrimmage, it was he who digs up the most toys wins.

At the other end, the waste end, we also believed the Earth to be limitless, always ready to digest and dilute our waste. We could toss as much garbage as we wanted into the surf, and it would never float back to shore.

"Economies are like ecosystems," says Allenby. "Both systems take in energy and materials and transform them into products. The problem is that our economy performs a linear transformation, whereas nature's is cyclic." We're like the juggler who takes a set of bowling pins, tosses them in the air once, then throws them out, reaching for a new set. Life, on the other hand, juggles one set of pins and cycles them continually. A leaf falls to the forest floor only to be recycled in the bodies of microbes and returned to the soil water, where it is reabsorbed by the tree to make new leaves. Nothing is wasted, and the whole show runs on ambient solar energy.

Industrial ecology asks the simple question, what if this closed-loop, sun-driven biology were to become our *modus operandi*? What if our economy were to deliberately look and *function* like the natural world in which it is embedded? Wouldn't we be more likely to be accepted and sustained by the natural world over time? This, in a nutshell, is the dream of industrial ecology.

The idea itself is not new; similar thoughts have been percolating in the environmental literature since the sixties. What is new is that some of the staunchest proponents of this philosophy are swiveling in executive chairs at the world's largest companies. Bob Laudise explains how industry managers began to green around the edges during the 1990s, and how conscious emulation of natural systems became the hottest business shibboleth since Total Quality Management.

THE GREENING OF INDUSTRY

W. Edwards Deming (the father of Total Quality Management) taught us to look for and fix the root causes of problems. In the long run, he said, quick fixes leak and need shoring up. TQM adherents like Braden Allenby realized that pollution was not the root cause of our environmental crisis; fantasy was. We had begun telling ourselves a dangerous fairy tale that went something like this: The Earth, put here for our use, is a limitless provider of resources and will clean up our messes for free. We treated raw materials as if they were essentially free—you paid for access to them and you paid to remove them, but you paid nothing for the leaching slag heaps or the fact that you were depleting another generation's resource stocks. Waste was released to oceans, rivers, land, and air, with no recompense for the Earth's free services.

A pricing scheme that ignored environmental costs was a silent perpetuator of this ruse. Because the economy put no price tag on resource drawdowns or on pollution, it gave no incentive to extract sustainably, process cleanly, or optimize use. As a result, Laudise says, "We made dumb materials choices, dumb process choices, and when it came to waste, we blithely elected to emit it and forget it." For a long time, like adolescents who think they are immortal, we acted as if we had some sort of magic shield against the consequences of our plundering and polluting.

As for activities that caused pollution, they were all but lionized in the name of "progress." I have a 1930s rubber stamp that has a downright heroic-looking set of smokestacks belching for all they are worth. The idea was to place this at the top of your letterhead to symbolize your own prosperity. When I told Laudise about the stamp, he showed me some equally glowing "factory cards" that were collected and exchanged like baseball cards. Evidently there was no greater source of pride than to have the "World's Largest Fertilizer Factory" in your town. Enabled by the economy and blind to the dangers, we climbed to a great height of delusion, and became more determined than ever to keep those smokestacks waving.

In the 1960s and 1970s, *bang!* The first warning shots were fired about the health effects of environmental pollutants, with some of the most ringing salvos coming from the pen of Rachel Carson. The environmental movement woke with a start and surged forward to win many legislative victories. It was the beginning of the "command and control" laws, which directed industry to muzzle its smokestacks and cauterize the hemorrhaging at the ends of its pipes. Like all rules exerted from above, however, command-and-control laws were just begging to be circumvented. Companies quickly hired squadrons of lawyers to perfect the art of minimal compliance. By the indulgent eighties, denial was back in style, and corporations routinely lobbied to reverse environmental regulations or, failing that, found ways to wriggle under them. It gave stockholders and consumers one last, short-lived hurrah.

Instead of fading from fashion, however, federal regulations kept growing in number and severity, doubling between 1970 and 1990. Toward the end of the eighties, the original laws moved into their more stringent phases, loopholes closed, and states and local governments stepped up to the plate with their own antipollution laws. As Laudise showed in one of his viewgraphs, corporations faced a relentlessly climbing slope of regulatory red tape.

With each step toward compliance, costs ratcheted up as well. According to the National Renewable Energy Laboratory, U.S. industry is spending $70 billion per year treating and disposing of its wastes. Even these economic penalties failed to sober all the partiers, however. What really sent corporate America back to the drawing board in the nineties was the greening of its customers.

Ecologist Paul Ehrlich says that we are not hardwired genetically to respond to long-term dangers—it takes a saber-toothed tiger roaring at the cave mouth for us to jump out of our skins. These days, the environmental sabertooth licks its chops on our televisions, in our newspapers, in our wells, and on our beaches, and our skin is finally beginning to crawl.

One particularly memorable cat in the cave mouth came in 1987 when a barge laden with 3,186 tons of commercial garbage left Islip, Long Island, and spent the next six months looking for a place to dump its load. No one wanted it, and the bloated barge kept rising up over the horizon, proving once and for all that the world is not flat—it has no convenient edge over which we can shove all our disposables. The next year, lest we think that was an isolated occurrence, the cargo ship *Khian Sea* left Philadelphia with 15,000 tons of toxic incinerator ash and roamed for two years before it finally dumped its waste in an "undisclosed" location. The world had never looked so small or over-burdened. We followed the barges' journeys with nauseated fascination, the way we had watched the senseless violence of television wars and on-camera assassinations. Now it was the Earth's turn.

The images kept on coming. The cows of Chernobyl sickening, rivers in the Ukraine catching on fire, the smothering oil fires of the Persian Gulf, a ship leaking death into Prince William Sound, syringes surging around the ankles of New Jersey swimmers. The soundtrack to all this was the Cassandra choruses of scientists warning of an ozone hole twice the size of Europe, a smoggy Arctic Haze thousands of miles from the nearest city, rafts of amphibians blinking out like warning lights, and strange reproductive deformities afflicting dozens of wildlife species.

All the while, our population mushroomed, sending industrial fallout to each corner of the Earth. Europe's trees began to weaken, the deserts marched, the rain forests shrank, and the wetlands dried out, exhaling their petrified cache of carbon in "greenhouse gas" form. Even the weather seemed to have gone mad, as if Gaia were sneezing us out of her system. By now, people had had enough—enough Love Canals, enough Bhopals, enough Cancer Alleys, enough Summers of 1988.

These days, citizens welcome dirty industries into their backyards about as readily as they'd welcome the *Khian Sea* to their bathtub. Thanks to Community-Right-to-Know legislation, newspapers carry the records of emissions of neighboring businesses, opening them to community shame. In editorials across the nation, smokestacks are referred to as smoking guns, firing the equivalent of shrapnel into our lungs. People are making a commitment to personally "do something about the environment," making surprise best-sellers out of books such as *50 Ways to Save the Earth*. Consumers are also voting at the cash register, weighing in against dolphin-abusive tuna-netting practices and for organic agriculture. Overnight, it seems, people who litter or refuse to recycle have begun to seem, to say the least, unsavory.

And it's not just happening in yuppie America. Here and abroad, surveys have shown that an astounding percentage of people are concerned about the environment and are willing to change their lifestyles. A 1992 George Gallup Health of the Planet Survey showed that between 40 and 80 percent of the respondents from twenty-two countries are already "avoiding the use of products that harm the environment."

The tide has definitely turned. Soil loss, water poisoning, and air contamination, little more than background static up until now, have suddenly become *information*. The economy, a beast whose senses are tuned to customers' changing moods, is beginning to twitch. And a worried industry, concerned about covering its bottom line, is headed in droves to seminars like Laudise's.

Laudise speaks loudly and with punch, like a coach talking strategy to his team before going out for the second half. "OK. What we've realized is that despite all the happy consequences of industrialization—medical miracles and the common man being able to tune in the Philharmonic and all that—we can't go on like this. The way we've been operating is illogical from a sustainability point of view."

Heresy, right? But as I looked around the room, every head was nodding. As he went on, I had to keep reminding myself that this wasn't a Sierra Club meeting. It was a corporate strategy session, and Laudise was talking tough love. "There are three reasons for greening up your act: It's the right thing to do, it's the competitive thing to do, and you'll go to jail if you don't."

One way or another, corporate America and consuming America are starting to get the picture. We are realizing that there is nowhere to run, no edge of town where we can pile our wastes out of sight and out of mind. The world is a roundabout, and we are not immune to its laws, its boundary conditions.

At this point in history, our problem is not a shortage of raw materials (though that will come), it's that we've run smack against the limits of the Earth's resilience. As Tibbs says, "The natural environment is a brilliantly ingenious and adaptive system, but there are undoubtedly limits to its ability to absorb vastly increased flows of even naturally abundant chemicals and remain the friendly place we call home." Our manufacturing output is now twice as high as it was in 1970, and many products that did not even exist twenty-five years ago are being manufactured in mass quantities. That's a lot of barge trips to nowhere.

We can go one of two ways, Laudise told the crowd. We can either crash to a subsistence population level, with all the horrors of a second Dark Ages, or we can find a way to provide a quality life for a stable population (assuming we can achieve that) without over-stressing nature's filters. In short, if we play our cards right, we could pull off a "soft landing." More nods. Count industry in.

Suddenly, the green path has become the most intelligent and maybe even the most profitable way out of the mess for the corporation tuned to survival. Al Gore dangles the bait in his book *Earth in the Balance*: "The global market for environmental goods and services is approximately $300 billion and is expected to grow to $400 to 500 billion by the beginning of the next century. If one includes recent estimates for investments in energy infrastructure in developing countries, this figure grows to more than $1 trillion by the end of the decade." Sure it's self-interest—companies want to get ahead of the green wave so they can surf it, not be crushed by it. And they sure as heck want to get to shore before their competitors do. The feeling seems to be, if the environment gets cleaned up along the way, that's great too.

To me it doesn't really matter why industry wants to change its colors. The important point, though it's not always public knowledge, is that many companies do want to change. Even as they are pressuring Congress to relax environmental regulations, they are meeting to find out how to make Earth-friendly products in Earth-friendly ways.

This means an enormous segment of the public—stockholders, workers, managers, consumers—are out shopping for ideas that will work: a new way to think, a new paradigm that will guide our hand as we dismantle the economy we have so feverishly erected and re-place it with something that will sustain. As Einstein said, "The significant problems we face cannot be solved by the same level of thinking that created them." People like Laudise and Tibbs pack the house because they have a simple, compelling idea that hails from a group of people that industry traditionally hasn't consulted.

You won't find their books in the airport business bookstalls. They don't come from Harvard Business School or California think tanks or Japanese productivity institutes. The consultants of the nineties come blinking into the artificial lights of corporate conference rooms fresh from butterfly counts, gorilla watches, and bird bandings. As they put on their first carousel of slides—coral reefs, redwood forests, prairies, and steppes—even E. F. Hutton is listening. This is what's so amazing to me. In the most unlikely and promising cross-fertilization of our times, the Birkenstocks are teaching the suits.

SURVIVING IN PLACE:
EMULATING NATURE'S ECONOMICS

William Cooper wonders what an old fish squeezer like him is doing on the *Journal of Urban Ecology*'s masthead, or the National Academy of Science's panel to investigate the building of six hundred supersonic transport planes. A fish biologist by training, Cooper has cultivated a multi-octave range of specialties, thriving in the tidal pool between disciplines that is home to good biomimics.

In addition to teaching zoology at Michigan State University, Cooper is an adjunct professor in marine sciences in Virginia, and civil, environmental, and mineral engineering in Michigan and Minnesota. He's chaired a department and seven advisory boards and is now on the editorial board of four journals. In fact, from the looks of his vita, you'd be hard-pressed to find a global change, waste management, or environmental risk board that Cooper has *not* served on. In his spare time, he works for the Brookings Institution, giving about thirty-five seminars a year to policy makers who are about to sail or sink important legislation.

Despite this heavyweight influence, Cooper is a surprisingly self-effacing, plain talker with a grounded sense of the absurd. I laughed a lot when I talked with him, and I imagine his students enjoy the boomerang rides he calls lectures.

A decade before it was fashionable, Cooper tells me, he wandered out of the Zoology Department at the University of Michigan and began to teach a class in ecological systems to engineers. When Braden Allenby heard about that, he invited Cooper to a 1992 Woods Hole meeting to talk about a newly birthed concept called industrial ecology. "I was the only biologist in the room," Cooper recalls.

What he told Allenby and the other business thinkers was good news. The natural world is full of models for a more sustainable economic system—prairies, coral reefs, oak-hickory forests, old-growth redwood and Douglas-fir forests, and more. These mature ecosystems do everything we want to do. They self-organize into a diverse and integrated community of organisms with a common purpose—to maintain their presence in one place, make the most of what is available, and endure over the long haul.

But he also told them some bad news. We are nothing like the equilibrium organisms we want to emulate. Right now, we are occupying a niche that is also found in the natural world—that of opportunists, concentrating on growth and throughput (how fast raw materials can be turned into products) without giving much thought to efficiency. We're acting as if we're only passing through, taking advantage of the plenty and then moving on.

Opportunists are the weeds in a farmer's newly turned field, the bacteria in a Tupperware of leftovers, or the mice in a catless barn. These communities, called Type I systems, spring up to take advantage of abundant resources. They typically use resources as quickly as they can, turning them into adult bodies and then into numerous, small offspring—thousands of insect eggs, for instance. The idea behind this rapid growth strategy is to grow your population, maximize throughput of materials, and then head for the next horn of plenty, with no time for recycling or efficiency. Sound familiar?

"The Industrial Revolution was the equivalent of throwing a handful of flour beetles into a fresh bin of clean, sifted flour," Allenby told me. We suddenly had unlimited resources, and like any opportunistic system, we went hog wild, with one important difference. Unlike flour beetles, who can eat and be merry and then move on to another bin of flour, we are in a finite container called Earth. To get a grim foreshadowing of our predicament, put a screen atop the flour bin so the beetles can't get out to find their next cornucopia.

The screened-in beetles will eat and reproduce, filling the bin with beetle bodies. Because their system is so simple, there is no decay segment of society, no janitorial species to clean up the corpses and convert them back into food. That means that once the flour gets turned into flour-beetle bodies, those nutrients are locked away from the increasingly hungry population. It's like our economy turning the last of our raw materials into products, with no mechanism for recycling those products.

Living space quickly becomes scarce as well. As the population reaches the peak of its classic sigmoid curve, the madding crowd begins to get in one another's way. Antennas are locking, beetles are munching on the offspring of other beetles, and copulating beetles are being interrupted by a third and a fourth before they can mate. Within days, survival rates teeter, births stall, and the population comes crashing to a "hard landing."

"It's not that these linear, Type I systems are categorically bad," says Bill Cooper. "That's a human judgment." If it weren't for Type I systems, the Earth's scars wouldn't heal. Annuals come in when soils have been disturbed—after fire, windfall, plowing, or plague. They carpet the ground, gobbling newly exposed nutrients and fertilizing the soil with their wastes, setting the stage for the grand conga dance called succession: Flower field turns to shrub field turns to forest. Though their moment in the sun is short, Type I pioneers can always find a new patch of disturbance somewhere, even in little gaps that are created after a tree falls. This slightly offbeat pulsing of decay and repair in many patches is what helps the community retain its stability.

But the strategy of ragweed, fireweed, and crabgrass doesn't work everywhere. It's only appropriate at the start-up stage of succession, when plenty of sunlight and soil nutrients are still available. Once the scene begins to crowd, and the pie of sun and water and nutrients is divided among more takers, the Type II strategy wins out.

The Type II system consists of perennial berry bushes and woody seedlings that move into the field. They are there for the longer haul. Unlike Type I species, they won't spend their energy on making millions of seeds. Instead, they'll make a few seeds and funnel the rest of the energy into hardy roots and sturdy stems that will see them through winter. In the springtime, their prudence will pay off—they'll rebound from their roots and reach quickly for the sun, outpacing and eclipsing the Type I annuals.

At the very end of the conga line are those species that take this patience strategy to the extreme, showing even more loyalty to place. Type III species (the ones that will inherit the site and remain dominant until the next big disturbance) do more with less. They are designed to stay on the land in a state of relative equilibrium, taking out no more than they put in.

Masters of efficiency, Type III species don't have to go looking for sunlight. Their seedlings can tolerate their parents' shade, so wave after wave of the same species can grow up here. Biologists call these species K-selected. They have larger and fewer offspring, which have longer and more complex lives. They live in elaborate synergy with the species around them, and put their energy into optimizing these relationships. Together, the mesh of life juggles materials endlessly. Virtually no wastes leach away, and the only energy imported is that of the sun. By the time a mature forest like this closes ranks, pioneer species are long gone, off to their next sunny fortune—a fire scar in a forest, a gap from a wind-torn tree, the crack in your driveway.

Type I species are the rolling stones of the world, colonizing rather than learning to close the loops. The reason the footloose strategy works for them, says Cooper, is that new opportunities are always opening up. Back before our world was full, when we still had somewhere else to go, the Type I strategy looked like a good way to stay one step ahead of reality. These days, when we've gone everywhere there is to go, we have to find a different kind of plenty, not by jumping off to another planet but by closing the loops here on this one.

BECOMING MORE LIKE A REDWOOD
THAN A RAGWEED

Now that our rootball has grown to fill the world, we realize: *We have to learn to be self-renewing right where we are.* What we're talking about is changing our very niche, our profession in the ecosystem. Cooper says it won't do to just tweak the current system and hope that we'll evolve, just as a common ragweed or fireweed could not be expected to evolve into a redwood. Instead, we must replace portions of our Type I economy with portions of a Type III economy until the whole thing mirrors the natural world.

The gurus for this kind of niche-shift will be people who have studied the places we want to go. Systems ecologists like Howard T. Odum have studied the food chains in a prairie or estuary or bottomland and then drawn diagrams of energy flows and fluxes. If you didn't know better, you would think they were flow diagrams of a manufacturing process, complete with kilocalories per unit of "product" produced. Of all biologists, these folks come closest to speaking the language of process engineers.

When the flow chart of a developing Type I system is compared with that of a mature Type III system, some stark differences reveal themselves. This comparison table, first reproduced in a paper by Allenby and Cooper, represents decades of work by systems ecologists like Odum. Many of these concepts will appear in the upcoming discussions.

ECOLOGICAL SUCCESSION

Ecosystem Attributes	Developing Stages (Type I)	Mature Stages (Type III)
Food chain	Linear	Weblike
Species diversity	Low	High
Body size	Small	Large
Life cycles	Short, simple	Long, complex
Growth strategy (how to multiply)	Emphasis on rapid growth (r-selection)	Emphasis on feedback control (K-selection)
Production (body mass and offspring)	Quantity	Quality
Internal symbiosis (cooperative relationships)	Undeveloped	Developed
Nutrient conservation (closed-loop cycling)	Poor	Good
Pattern diversity (vertical canopy layers and horizontal patchiness)	Simple	Complex
Biochemical diversity (such as plant-herbivore "arms races")	Low	High
Niche specializations (jobs in the ecosystem)	Broad	Narrow
Mineral cycles	Open	Closed
Nutrient exchange rate between organisms and environment	Fast	Slow
Role of detritus (dead organic matter) in nutrient regeneration	Unimportant	Important
Inorganic nutrients (minerals such as iron)	Extrabiotic	Intrabiotic
Total organic matter (nutrients tied up in biomass)	Small	Large

Ecosystem Attributes	Developing Stages (Type I)	Mature Stages (Type III)
Stability (resistance to external perturbation)	Poor	Good
Entropy (energy lost)	High	Low
Information (feedback loops)	Low	High

Adapted from Braden R. Allenby and William E. Cooper, "Understanding Industrial Ecology from a Biological Systems Perspective," *Total Quality Environmental Management*, Spring 1994, pp. 343–354.

You can read this chart as a list of challenges or lessons—column two is where we are now, the ragweed stage, and column three is the redwood stage, the blueprint for our future survival. Though the two appear to be worlds apart, industrial ecologists are quick to note that the ragweed economy and the redwood ecosystem are both complex systems, and as such, they have much in common.

Complex systems—such as a wildfire, a storm pattern, or a waterfall—are not "run" by anyone in particular, but are instead controlled by countless individual interactions that occur inside the system. Every day, for instance, customers in hundreds of countries make decisions to buy or not to buy, and those decisions in turn affect the price of beans and stocks. In the same way, countless interactions in a natural system—eating or being eaten, for instance—weave together to define the community. Just as the invisible hand of the marketplace determines whether a company lives or dies, so natural selection works from within to shape the nature of life.

Over billions of years, natural selection has come up with winning strategies adopted by all complex, mature ecosystems. The strategies in the following list are tried-and-true approaches to the mystery of surviving in place. Think of them as the ten commandments of the redwood clan. Organisms in a mature ecosystem:

1. Use waste as a resource
2. Diversify and cooperate to fully use the habitat
3. Gather and use energy efficiently
4. Optimize rather than maximize
5. Use materials sparingly
6. Don't foul their nests
7. Don't draw down resources

8. Remain in balance with the biosphere
9. Run on information
10. Shop locally

If we agree there's merit to trying to emulate these approaches, it's easy to see that our economy, since it is also a complex system, has more than a snowball's chance of actually being able to operate and survive this way. This hope is what motivates industrial ecologists to get up every morning and work to shift our niche.

Living the Lessons

Though they know it won't happen all at once, the Allenbys and Tibbses of the world want to move us toward a future in which industry runs on sunlight (or a similar renewable nonpolluting source), doesn't "overdraw" natural resources or foul its own nest, sees nothing as waste, is cooperative and diversified, and does more with less through ingenious, high-quality, information-rich design of products and processes. In short, they envision an industry that is more like a closed-loop redwood forest than my front lawn.

As you will see in the comparisons that follow, our culture is taking some first tentative steps down this "path of no regrets." Right-thinking companies, shaped by their own form of natural selection, are already experimenting with the approaches you will find here, trying to mimic the successes of redwood communities. If any company or national economy is successful in applying all ten lessons, it could master a trick that's as old as the first bacteria: life creating conditions conducive to life.

1. Use Waste as a Resource.
One of the key lessons from systems ecology is that as a system puts on more biomass (total living weight), it needs more recycling loops to keep it from collapsing. A forest is more complex than a weed field—shrubs and trees and vines and mosses and lichen and squirrels and porcupines and bark beetles extend upward and outward, filling every nook and cranny with life. If all that biomass kept withdrawing nutrients from the environment with no way of recouping from within, it would quickly suck its surroundings dry.

Instead, the mature community becomes more and more self-contained. Rather than exchanging nutrients and minerals with the outside environment at a high rate, it circulates what it needs within its pool of sprouting, dying, and decaying organic matter. The reason the cycle works so smoothly is that there are no holes in the organizational chart—a diverse assembly of producers, consumers, and decomposers have evolved to play their parts in closing the loops so resources won't be lost. All waste is food, and everybody winds up reincarnated inside somebody else. The only thing the community imports in any appreciable amount is energy in the form of sunlight, and the only thing it exports is the byproduct of its energy use, heat.

Using Waste as a Resource: The Lessons Learned

If anybody's growing biomass, it's us. To keep our system from collapsing on itself, industrial ecologists are attempting to build a "no-waste economy." Instead of a linear production system, which binges on virgin raw materials and spews out unusable waste, they envision a web of closed loops in which a minimum of raw materials comes in the door, and very little waste escapes. The first examples of this no-waste economy are collections of companies clustered in an eco-park and connected in a food chain, with each firm's offal going next door to become the other firm's raw material or fuel.

In Denmark, the town of Kalundborg has the world's most elaborate prototype of an ecopark. Four companies are collocated, and all of them are linked, dependent on one another for resources or energy. The Asnaesverket Power Company pipes some of its waste steam to power the engines of two companies: the Statoil Refinery and Novo Nordisk (a pharmaceutical plant). Another pipeline delivers the remaining waste steam to heat thirty-five hundred homes in the town, eliminating the need for oil furnaces. The power plant also delivers its cooling water, now toasty warm, to fifty-seven ponds' worth of fish. The fish revel in the warm water, and the fish farm produces 250 tons of sea trout and turbot each year.

Waste steam from the power company is used by Novo Nordisk to heat the fermentation tanks that produce insulin and enzymes. This process in turn creates 700,000 tons of nitrogen-rich slurry a year, which used to be dumped into the fjord. Now, Novo bequeaths it free to nearby farmers—a pipeline delivers the fertilizer to the growing plants, which are in turn harvested to feed the bacteria in the fermentation tanks.

Meanwhile, back at the Statoil Refinery, waste gas that used to go up a smokestack is now purified. Some is used internally as fuel, some is piped to the power company, and the rest goes to Gyproc, the wallboard maker next door. The sulfur squeezed from the gas during purification is loaded onto trucks and sent to Kemira, a company that produces sulfuric acid. The power company also squeezes sulfur from its emissions, but it converts most of it to calcium sulfate (industrial gypsum), which it sells to Gyproc for wallboard.

Although Kalundborg is a cozy collocation, industries need not be geographically close to operate in a food web as long as they are connected by information and a mutual desire to use waste. Already, some companies are designing their processes so that any waste that falls on the production-room floor is valuable and can be used by someone else. In this game of "designed offal," a process with lots of waste, as long as it's "wanted waste," may be better than one with a small amount of waste that must be landfilled or burned. As author Daniel Chiras says, more companies are recognizing that "technologies that produce byproducts society cannot absorb are essentially failed technologies."

So far, we've talked about recycling within the confines of one manufacturing plant or within a circle of companies. But what happens when a product leaves the manufacturer's gates and passes to the consumer and finally to the trash can? Right now, a product visits one of two fates at the end of its useful life. It can be dissipated to the environment (buried in a landfill or incinerated), or it can be recaptured through recycling or reuse. The closed-loop dream of industrial ecology won't be complete until all products that are sent out into the world are folded back into the system.

Traditionally, manufacturers haven't had to worry about what happens to a product after it leaves their gates. But that is starting to change, thanks to laws now in the wings in Europe (and headed for the United States) that will require companies to take back their durable goods such as refrigerators, washers, and cars at the end of their useful lives. In Germany, the take-back laws start with the initial sale. Companies must take back all their packaging or hire middlemen to do the packaging recycling for them. Take-back laws mean that manufacturers who have been saying, "This product can be recycled," must now say, "*We* recycle our products and packaging."

When the onus shifts in this way, it's suddenly in the company's best interest to design a product that will either last a good long time or come apart easily for recycling or reuse. Refrigerators and cars will be assembled using easy-open snaps instead of glued-together joints, and for recyclability, each part will be made of one material instead of twenty. Even simple things, like the snack bags for potato chips, will be streamlined. Today's bags, which have nine thin layers made of *seven* different materials, will no doubt be replaced by one material that can preserve freshness and can easily be remade into a new bag. And that bag will most certainly be marked with a universal material code, making it easier for the companies charged with take-back to recycle and refurbish them.

As Allenby explained, take-back laws are a change in the market environment, and the companies that want to survive in that habitat are already evolving. BMW's new sports car, for instance, can be broken down in twenty minutes on an "unassembly" line. ("I wouldn't want to leave one of these on the streets of New York," kids Laudise as he shows me before-and-after pictures.)

Refurbishment is another key to giving products a longer life in the marketplace. Instead of buying a new computer case each time you want to upgrade, you'll most likely buy the snappy new module and plug it into your original case. When you do hand over your old behemoth, it may be "mined" for parts which will be refurbished and show up again in new machines. "Asset recovery" is what Xerox calls it. The parts stripping and refurbishing program for its copiers saves the company $200 million annually.

The Canadian arm of Black & Decker has started a recycling system for its rechargeable appliances, hoping to reduce contamination and waste from nickel-cadmium rechargeable batteries. Customers have the choice of either having the rechargeable batteries replaced or leaving the products with a local distributor for recycling. As an incentive to bring the item in, customers who do so are eligible for a five-dollar rebate toward their next Black & Decker product. So far, 127 fewer tons of waste (including 21 fewer tons of nickel-cadmium batteries) were landfilled in Ontario, where the program has been piloted. Black & Decker also benefits from future sales that the rebate system encourages.

Canon, in response to worldwide demand for recycling, is also inviting customers to mail in their old ink cartridges from printers and copiers. The postage is paid by the company, and for each one mailed in, Canon sends a five-dollar donation to either the National Wildlife Federation or The Nature Conservancy.

Businesses that have been in the game for a while report that being green is good for profits. Anita Roddick's Body Shop has made a fortune on the concept of refilling customers' containers of cosmetics and toiletries to cut down on packaging waste. Déjà Shoe (my candidate for best green name) makes old tires into shoes, claiming it's better to wear them than burn them. Patagonia does the same for pop bottles, polishing its already verdant image by offering the first guilt-free polar-fleece jackets. With waste-recovery successes like these, suggests Allenby, we might as well stop calling it waste.

2. Diversify and Cooperate to Fully Use the Habitat.

The more we learn about nature's resource allotment strategies, the more it looks like Tennyson had it only half right when he said nature was "red in tooth and claw." In mature ecosystems, cooperation seems to be just as important as competition. Using cooperative strategies, organisms spread out into noncompeting niches and basically clean up every crumb before it even falls off the table. This diversity of niches creates a dynamic stability; if one organism drops out of the network, there's usually a backup, allowing the web to stay whole.

Even when individuals within a species share a niche, there are "agreements" about resource allotment. Animals will claim territories, for instance, or feed at different times of day to avoid overlapping with their counterparts. As a result, the spoils of their habitat are divvied up so that whole gaggles, herds, troops, and coveys can be supported by the same piece of land without constant energy-draining fights. This "peaceful coexistence," writes ecologist Paul Colinvaux, is inherently cooperative, though it may not be a conscious pact as it is with humans.

More overt forms of cooperation can be seen in the partnerships that some animals form for mutual benefit. The classic example is the goby fish that picks parasites from the teeth and gills of the Nassau grouper fish. In return for this cleaning service, the grouper resists eating the tiny goby and actually protects it from other predators. Noisy oxbirds also perform a service, alerting hippos to interlopers in return for being allowed to dine on ticks embedded in the hippo's skin. Lichen represent a more permanent arrangement between two species: Algae and fungi move in together, one harvesting solar energy, the other providing a safe support structure. What emerges when you combine talents like these is synergy—a sustainable system far greater than the sum of the parts.

Lynn Margulis, co-author of the Gaia hypothesis (the idea that the Earth is self-regulating, like a living organism), believes that symbiosis is not confined to a few oddball species, but is in fact essential to all evolution. According to the endosymbiotic hypothesis, which she has written about extensively, a large leapfrogging of progress occurred billions of years ago when two species joined forces. A bacteria that couldn't manufacture its own food engulfed another bacteria that could photosynthesize. Instead of being killed, the green "boarder" stayed on, and has stayed on to this day. In fact, says Margulis, the successors to these symbionts are the chloroplasts that exist in all green plants. Another symbiotic story can be seen in the oxygen-breathing, energy-producing organelles in our cells called mitochondria. Proponents of this hypothesis, which is widely accepted, postulate that these mitochondria were free-ranging bacteria at one time, which explains why they still have their own set of DNA.

If the endosymbiotic hypothesis is true, then every cell in our body is a symbiotic creature. When these symbionts gather in great herds, they form organs and organisms. In fact, writ large, the theory goes like this: Our body is actually an aggregate of single-cell creatures that have formed a giant multicellular assembly. In short, we are a colony—a single organism composed of many—and a living proof of the power of cooperation.

Diversifying and Cooperating: The Lessons Learned

Anyone who has collected green bottles for several months only to hear "Sorry, we can't recycle green glass—no markets" knows the frustration of the web that has holes. The more pathways we have for feeding off each other in the industrial ecosystem, the more loops will be closed and the less waste will be lost from the system.

Right now, within the linear extract-and-dump model, the niches—the jobs within the web—are not all in place. As the Japanese industrial ecologist Michiyuki Uenohara says, we have plenty of "arteries"—ways for products to flow from the heart of manufacturers into the body of the economy—but we need "veins" as well, ways to return the products so that their materials can be purified and reused. As part of Japan's Ecofactory Initiative, restoration factories are being built nationwide to refurbish or recycle products at the end of their life.

The Japanese are also building a form of cooperation into the design phase of their product development. In this strategy, the competitive whistle doesn't blow until marketing begins. Prior to marketing, companies participate in common goals like Design for Disassembly. This notion of precompetitive cooperation is also showing up in the United States, the most notable example being the Vehicle Recycling Partnership of Chrysler, Ford, and General Motors. Putting aside their normally fierce competition, companies like these are working through trade associations, special alliances, and "virtual firms" to come up with common labeling and materials standards which will allow them to reuse each other's parts. This kind of alliance-building is to be expected in an emerging Type III economy. The more veins and arteries you add to a system, the more complex it becomes, and the more cooperation you need for proper functioning.

One day, say industrial ecologists, the town that has no takers for green glass will be seen as a town with a niche unfilled, an opportunity that won't stay open for long. In an economy where veins and arteries are equally profitable, entrepreneurs will consciously work to sew up the loose ends of resource use and reuse. The result: a web without holes that looks and behaves more like a mature community.

3. Gather and Use Energy Efficiently.

Not everything needed by industry can be recycled, however. Even in a natural system, only nutrients and minerals can be circulated through the diverse connections of an ecosystem; energy cannot. In a salute to the Second Law of Thermodynamics, energy is converted to heat in the process of doing work, and is therefore unavailable to do more work. As a result, the energy that runs the juggler's art must be continually imported into the system.

In nearly every community (except sulfur-based "vent" communities on the ocean floor), the purchasing agents for energy are photosynthesizers—green plants, blue-green algae, and certain bacteria. They siphon their radiant energy from a nuclear fusion occurring 93 million miles away (the sun) and transform it into the chemical bonds of sugars and carbohydrates. Though they use only about 2 percent of the sunlight that reaches the Earth, they make the most of it, achieving an astounding 95 percent quantum efficiency. (That means that for every 100 photons of light captured by the leaf's reaction center, 95 are funneled into bond making.)

Next time you are in a leafy mature forest, take time to marvel at nature's efficient solar-collector array. Leaves are positioned relative to one another to maximize exposure, and like miniblinds, some actually tilt and swivel as the sun traces the sky. This efficient process collects energy for all living beings, and sets the ceiling for what an ecosystem can aspire to be.

The carrying capacity of the land has everything to do with how much energy there is to go around. After plants use their energy booty for growing and reproducing, only 10 percent is available to the next food chain level, the herbivores. Only 10 percent of that 10 percent is available to carnivores, and so on. That's why, as ecologist Paul Colinvaux says, "big fierce animals are rare," and why plants usually constitute most of the biomass (total living weight) in terrestrial ecosystems. The pyramid of life is quite literally an energy distribution chart, a record of the sun's movement through the system.

When you're only one of many species competing for a slice of the sun's energy, you can't afford to be capricious in your use of energy. That's why animals travel a minimum distance to get what they need, and time their activities to maximize their rewards and minimize energy costs. Plants send roots only as far as they need them and don't try to "tough it out" where the soil or water levels are wrong for them. Both animals and plants doggedly protect what they secure. The Southwest's collared peccary hoards its hard-won water (even its urine is crystalline), while the sugar maple of the North drops its leaves to seal off water loss during the winter. These energy-saving devices are not coincidence; those who overuse or squander energy are eventually edited out of the gene pool.

Thrift, on the other hand, pays handsomely. Even in the manufacture of bones and skin and shells and webs, organisms have evolved ways to work smarter, not harder. The use of enzymes for catalyzing or speeding up chemical reactions is a perfect example. A good enzyme can accelerate a chemical reaction by 10^{10} times (1 with 10 zeros after it). Without such acceleration, a process that takes five seconds, such as reading this sentence, would take fifteen hundred years. Biological catalysts also allow nature to manufacture benignly; instead of using high heats and harsh chemicals to create or break bonds, nature manufactures at room temperature and in water. The physics of falling together and falling apart—the natural drive toward self-assembly—does all the work.

Gathering and Using Energy Efficiently: The Lessons Learned

We could learn a lot from plants. Ideally, we too should use an external, renewable source of energy, specifically *current* sunlight (solar, wind, tidal, and biodiesel forms of power all rely on current sunlight). As it is, we are using *ancient* sunlight that was trapped here on Earth in the bodies of Cretaceous plants and animals. Because these remains were compressed without oxygen, they never got the opportunity to decay. Now, when we burn these fossil remains as oil, coal, or natural gas, we complete the decay process all at once, exhaling the stored carbon into the atmosphere in large doses, violating the "no large fluxes" ecosystem lesson. Unfortunately, as long as these ancient sources are still cheap, our energy-addicted society appears determined to burn them all.

Renewable energy expert Amory Lovins believes that until we can make the shift to gathering current sunlight directly, the best strategy is to coax every last kilowatt out of the fuels we are using. Already, many industries have discovered the monetary benefits of tightening energy leaks with devices such as compact fluorescent lights, weather-tight building panels, and energy-sipping appliances. Du Pont has reduced energy use per pound of production 37 percent since 1973. It's expecting to shave another 15 percent during the 1990s. In the last twenty years, while Japan's economic activity has increased, its energy consumption has actually *decreased*. It attributes this reduction to the substitution of information—good ideas—in place of more energy.

Utility companies in this country are beginning to help consumers plug the leaks at the company's expense. In western Montana, for instance, my rural electric cooperative, which buys from Bonneville Power, paid two thirds of the cost to insulate my attic. It believes that by weatherproofing its customers' homes, it can help keep power demand below the level that would force it to build a new power station. Though it seems incongruous, Bonneville sells less electricity this way but makes just as much money, because it has eliminated construction costs of new plants from its budgets. Everybody wins, including the environment.

But what about the huge drain coming from energy-intensive manufacturing? Unlike the slow burn of organic engines (cells), we are always beating, heating, and treating our materials to form them the way we want, wielding high fluxes of energy that would never be tolerated in natural systems. If the dreams of biomimetic materials scientists are realized, high energies will no longer be synonymous with manufacturing. Instead, our processes will mimic those of spiders, abalones, blue mussels, and other organisms on an energy budget.

Lessons from natural systems can also help us decide *what* to use our energy for. As Amory Lovins says, "If I were to come back in fifty years and find that we had extremely efficient factories making napalm and throwaway beer cans, I'd be very disappointed, because it would mean that we hadn't addressed a parallel agenda of what's worth doing with all that energy." Natural systems use their energy to maximize diversity so they can be more efficient in terms of mineral and nutrient recycling. Perhaps we should reevaluate what we are maximizing (throughput) and take a look at optimizing instead.

4. Optimize Rather Than Maximize.

A field of annual plants is, like we are, pushing throughput. It's turning nutrients into biomass, and just as quickly, it's turning biomass over, releasing plants back to the system when they die at the end of the year. Next year, the plants start from scratch again, accumulating the nutrients they need to jump through the hoop of rapid growth.

In contrast, the mature system keeps the bulk of its materials and nutrients "on the stump"; instead of passing nutrients through to decay each year, most of the biomass stays put. In the early years, members of the plant community grow quickly (that's why tree rings are widest at the center of the tree). In later years, as more trees and vegetation come to share the space, the growing slows down, and the productivity per unit of biomass—the transformation rate of materials being made into products—slows down.

This journey to a mature system always follows the same pattern. The emphasis on maximizing throughput and offspring shifts to an emphasis on optimizing—closing nutrient and mineral flows, and making sure one or two offspring survive. In a mature mode, organisms are rewarded for being efficient and learning to do more with less. Those that survive are those that live within their means. Slowing down flow rates also leads to overall system stability. As Cooper says, "One of the reasons ecosystems are so resilient is that they aren't doing anything in a hurry. The slower the flow rates, the more you can modulate the controls without wild fluctuations." Being able to control the system is important; it means the whole community is able to change and adapt as the environment demands.

Optimizing, Not Maximizing: The Lessons Learned

Our industrial ecosystems are currently in arrested adolescence; they are still based on high rates of productivity and growth—a steady stream of materials moving as quickly as possible out of the Earth and into shiny new things. Eighty-five percent of manufactured items quickly become waste. In fact, when you add municipal and industrial waste together, every man, woman, and child in the United States produces twice his or her own weight in waste every day. Together, it's enough to fill two Louisiana Superdomes daily.

The lesson is to slow down the throughput of materials, emphasizing the quality rather than quantity of new things. Cooper says, "As the natural system matures, it redefines its concept of success. That's what fitness is all about. In today's economy, our definition of success is rapid growth—if you grow faster than your competitor, you win. In tomorrow's world, winning will mean being more competitive, doing more with less, and being more efficient than your competitor. Companies won't need to be as big—in fact, it might be more profitable to be small and produce high-quality products and services."

This trend toward optimizing rather than maximizing will reverse a well-established tide. The Industrial Revolution really got cooking when Fordist assembly-line manufacturing was invented. Items that had once been hand-crafted could now be mass-produced. While this led to affordability, it also led to cheapness in products, and ultimately to the disposable, ticky-tacky sameness that we are drowning in today.

In the 1960s, Japan launched a so-called Quality Revolution (based largely on efficiency expert Edwards Deming's ideas, which were initially ignored in this country). They proved it was possible to boost quality, productivity, and profitability at the same time. In the last decade or so, designers have spotted the quality trend in other countries as well—durable items, made with care and imbued with personality, are being increasingly favored over cheap, ubiquitous imitations. We can at least hope that this is a sign of a transition to a mature marketplace.

Another sign of maturity is the slow but increasing acceptance of "factory refurbished" products (e.g., rebuilt engines, factory-serviced stereos and computers). Rather than trash a model because a new one has appeared, it would be much better for the environment if we could see how long we could keep the existing model in the marketplace. This would shift the emphasis from manufacturing a new model every year to manufacturing longer-lived designs and creating subsidiary companies devoted to remanufacturing and upgrading. As Allenby says, "Our economic system is geared to the sale of many widgets. If we change that to the *maintenance* of many widgets, we change what we care about."

5. Use Materials Sparingly.

Organisms build for durability, but they don't overbuild. They fit form to function, building exactly what is needed, with the bare minimum of materials and fuss. Honeycombs are an example of a structure that encloses the maximum amount of space with the minimum amount of walling material. With their antennae alone, honeybees sculpt every six-sided room to within 2 percent of "specs," thus achieving strength without squandering wax. Bone is another example of form fitting function. Despite its light weight, bone is arranged in a design that resists breaking, even when stretched or compressed. The bones of birds epitomize this lesson—their strong, airily hollow skulls are what one engineer called "a poem in bone."

Organisms have also evolved to make the most of every design decision, by having one structure serve not just one but two or three functions. This constant adapting and reassessing of material use means that fewer devices have to be built for survival. Being good at this game gives organisms an edge, the difference between propagating your genes, or taking a one-way slide to hereditary oblivion.

Using Materials Sparingly: The Lessons Learned

"Green design" engineers, like equilibrium organisms, also love to do more with less. The current trend toward "dematerialization" allows companies to use less material to produce a lighter, smaller, sleeker product that performs many functions. Computers that fit in your palm and all-in-one fax/printer/copier/scanners are cases in point. Even heavy-duty products made of metal are getting thinner and stronger. Car bodies have shed about nine hundred pounds since 1975, and testosterone is no longer needed to crush a beer can with your bare hands. Creating a synergy between two types of materials—a composite—is another way to gain strength without adding bulk. Glass fibers woven through plastic make for stronger boat bodies, while carbon fibers in graphite give the Stealth bomber its edge.

The ultimate in dematerialization is a movement that may be described as "leasing as a way of life." Proponents of the so-called functional economy claim quite rightly that people don't want to own a heater, a refrigerator, or a TV; what they really want is heating, refrigeration, and entertainment. When they buy a CD, they want to hear music, not own a shiny disc.

Brad Allenby explained, "Imagine how things would change if the only physical objects you bought were those you wanted to own for sentimental or aesthetic reasons. Everything else around your house would be leased as a service. Various providers would be responsible for installing, maintaining, upgrading, and eventually replacing your appliances, your furniture, even your cookware."

Since the company would be responsible for uninterrupted service, the products it made would be reliable, heavy duty, and easy to repair and upgrade. "It would be like those old AT&T phones which were designed to last forty years," says Bob Laudise. "Back then, designers would trade off failure rates and service rates—they were responding to a different set of incentives. Now companies hope their product burns out so they can sell the consumer a new one."

In "leasing as a way of life," planned obsolescence would be, well . . . obsolete. Allenby tells me what an evening might be like in the functional economy: You drive home in your leased car, which was tuned up for you while you were at work. The mechanic came to your company parking lot, part of the service option that convinced you to renew your contract with this company. At home, you find that your well-built, energy-efficient refrigerator is keeping your foods even crisper than before. The service provider upgraded the coils last week so they could claim to have the most energy-efficient refrigeration on the market—something every company is struggling to provide these days.

You head to your leased stereo/TV/computer console and select some tunes from the digital collection of music you own rights to. When you bought the rights, you dialed into a digital server (a huge computer that holds all music archives) and downloaded the music onto your computer/player. There were no retail outlets, no CD jewel boxes, no packaging, no cash registers, no cardboard packing boxes piled in a Dumpster outside a neon-striped building.

While you're listening, you tell the computer to download the newspaper of your choice onto a thin portable reading tablet—or even better, you have your computer read it to you as you cook dinner on your leased range. After dinner, you hop on the Internet and request information about the latest generation of modems. You decide to upgrade your transmission speed and call in your order. Within seconds, the software upgrade arrives digitally through the wires, and your machine signals that the new modem speed is installed. No software store to go to, no packages to dispose of, no bulky manuals to clog your bookcase. I could get used to that.

The obvious question is, what happens to the companies that used to manufacture the CDs and other objects designed for obsolescence? What about the sales clerks in the software store? Allenby and his colleague Thomas Graedel, Distinguished Member of the Technical Staff at AT&T Bell Labs, recognize the dilemma. "A system that's geared for maximum-velocity throughput is quite different from one based on extending a product's life cycle or replacing products with services. We're going to have to decide which system we want."

Or, I think, we could wait for dwindling resources and overflowing landfills to decide for us.

6. Don't Foul Their Nests.

Organisms must eat, breathe, and sleep right in their manufacturing facility, their habitat; they can't afford to poison themselves. As a result, even poisonous snakes don't store their toxins in bulk; instead they create small batches only when needed. Nor do organisms use high heats, strong chemicals, or high pressures to manufacture as we do. They know that a high flux, or energy out of place, can contribute to nest fouling. Instead, organisms build their bodies using catalysts and self-assembly techniques, riding the free roller coaster of physics to put together adaptive materials. Moderation in energy and material use is the order of the day. By not stressing the supply lines or cleanup mechanisms in their environment, organisms win the right to keep on making a living there.

But life does more than simply keep its nest clean; it actually creates the conditions necessary for life. We are the only species that seems oblivious to this fact, and our insistence on polluting the lungs and filters of our world is evidence of our dense denial.

Not Fouling Our Nest: The Lessons Learned

It's easy to say, "Don't emit pollutants at rates greater than the Earth's capacity to assimilate them," but just how do we hold our industrial breath? Perhaps the best way to keep from fouling our air, water, and soil is to stop producing toxins, or abnormally high fluxes of any sort, up front. Industrial ecologists call it pollution prevention or precycling.

The Minnesota-based corporate giant, 3M, jumped on the pollution prevention bandwagon twenty years ago with an employee suggestion program called 3Ps (Pollution Prevention Pays). According to their own accounting, 3Ps has saved them an estimated $750 million and spared the Earth another 1.2 billion pounds of waste. All together, the company has adopted 4,350 cleaner-production projects in categories such as product reformulation, process modification, equipment redesign, recycling, and the recovery of waste materials for resale.

In each case, says 3M representative Jo Ann Broom, eliminating toxins from processing proved cheaper than cleaning up the toxins afterward. The first few years saw tremendous reductions in pollution, as companies changed procedures that were easy to change, referred to in the industry as "low-hanging fruit." Reaching beyond this point, like retrieving the topmost apples from a tree, may involve more of an effort. Nevertheless, 3M announced in 1988 that it intends to "cut all hazardous and non-hazardous releases to the air, water, and land by 90 percent and reduce the generation of waste 50 percent by the year 2000 (base year 1990). The ultimate goal is to reduce emissions to as close to zero as possible." Some other companies are following 3M's self-policing suit. Monsanto says it will reduce emissions of toxic chemicals by 90 percent by 1992 and reduce all waste by 70 percent by 1995. By 1993, Du Pont had already met its goal of reducing toxic air emissions by 60 percent from the base year of 1987 and was three quarters of the way toward the goal of reducing airborne carcinogens by 90 percent by the year 2000.

In the meantime, until we can completely eliminate or find a substitute for toxins, industrial ecologists are recommending we follow the "snake-venom law": manufacture chemicals in small doses where and when you need them, so you won't have to worry about storage or leakage. It's called "chemicals-on-demand," and industry's "venom glands" are small chemical generators built right into the assembly line. AT&T, for example, uses an on-site electrolysis machine that produces arsine (a dangerous gas) from its less harmful cousin, arsenic. Since the gas is produced right where it's needed, it saves AT&T the cost of transporting arsine (which is subject to dangerous, time-consuming, legally strict handling procedures) and averts the risk of spillage. Other extremely hazardous chemicals that would be good candidates for on-demand generators are vinyl chloride, methylisocyanate, phosgene, hydrazine, and ethylene chlorohydrin.

Another form of spillage is the "halo" of waste that occurs whenever chemicals such as paints or coatings are sprayed onto objects. Bob Laudise told me about a new application technique called molecular beam epitaxy that lays down extremely thin coats and directs the material where it should go and nowhere else. Costs to the Earth and to the company are reduced.

Another movement that is reducing waste is "just-in-time" manufacturing. In Japan, JIT factories are ringed by suppliers that are all connected by a computerized supply network. The suppliers make only what the factory needs on a hour-by-hour basis, so there is less warehouse storage and overproduction of goods. Levi-Strauss is trying the same technique, installing computers in its retail stores so it knows exactly how many jeans have sold and how many need to be made that day.

One last trend that would put us closer to making things the way nature does is the decentralization of production facilities. The most sensible place for this to occur would be in energy production. As Amory Lovins points out, we don't crowd all the dairy cows into one state and ship milk from there. Milk is perishable, so decentralized facilities make sense. Electricity, he argues, is perishable in its own way (it leaks through wires, and because of electrical resistance, it takes energy to move energy). It would make more sense for energy to be produced at small stations, or even on the rooftop of your home. The smaller the production load and the shorter the "commute," the less likely it is that large, nest-fouling fluxes or massive breakdowns will occur.

7. Don't Draw Down Resources.

Organisms in a mature ecosystem live on harvestable interest, not principal. The best predator, for instance, is the one that doesn't completely eliminate its prey. Likewise, the prudent parasite doesn't kill its host. Given the room, buffaloes will roam methodically rather than "nub down" their grassy prairies; giraffes will wander from acacia to acacia; and even voracious gorillas will move slowly through the jungle, allowing food plants to grow back in behind them. All have learned, through the wisdom of their genes, that gouging their growing stock is not a good idea.

Once again, the idea of organisms as mortal enemies locked in a zero-sum game just doesn't stand up to scrutiny. There are negative feedbacks in nature that keep organisms from completely chewing off the hand that feeds them. Namely, as food sources start to run out, they become harder to find, and the searching takes precious energy. Moving to an alternate food source is usually easier for the animal, and it allows the renewable stock to renew itself.

As far as nonrenewable resources like metals or minerals go, organisms don't use a whole lot of those to begin with, which may be a very big hint. The tiny helping of minerals taken up by organisms is replenished either through biological processes or through geological processes, such as uplifting, which brings buried minerals to the surface.

Not Drawing Down Resources: The Lessons Learned

Two corollaries to the lesson "Don't emit pollutants faster than the Earth can handle them" would have to be:

1. Don't use nonrenewable resources faster than you can develop substitutes.
2. Don't use renewable resources faster than they regenerate themselves.

At one time, our economy was primarily based on renewable materials—wood, natural fibers, plant-derived chemicals, and so on. One of our greatest missteps was to replace this economy with one based on *nonrenewables* such as oil, gas, coal, metals, and minerals. The law of sustainability says you should use nonrenewables at the same rate at which you are developing substitutes. But we are obviously using metals and minerals and fossil fuels faster than we are developing substitutes. If we are to leave any resources for our grandchildren, we should be recycling nonrenewables now, even if it means finding a way to "mine" landfills, where metals and minerals are often found in higher concentrations than they are in ore deposits!

The leak that will be toughest to seal is that of dissipative losses, those tiny bits of nonrenewables that are lost to the land, air, or water with each use. (For instance, your brake pads shower the road surface with material each time you stop.) Chemicals are especially prone to be dissipative; if they are not embodied in plastics, synthetic rubber, or synthetic fiber, they are likely to be in the use-and-lose category that includes coatings, pigments, pesticides, herbicides, germicides, preservatives, flocculants, antifreezes, explosives, propellants, fire retardants, reagents, detergents, fertilizers, fuels, and lubricants. Sealing up these slow leaks, and concentrating on recovery, may save virgin sources for generations to come.

Perhaps the best remedy of all is to find renewable substitutes for these nonrenewables. Recent talk about biopolymers, plastics from plants, and fuel from corn is evidence of a shift from rare and precious resources to ones that can conceivably be grown with the sun's help.

Not that a return to a renewable economy would be a total panacea. As Daniel Chiras warns, improperly managed timber cutting, farming, fishing, and ranching can result in severe erosion and marked decreases in the productive capacity of land and sea. The smart alternative is to take from the land only what will allow more to grow back. In forestry this is known as sustained yield, and the idea is to harvest only what has grown that year, so you are basically living on interest, not depleting the capital, of growing stock. The *capacity* for growing more is what must be protected. Unfortunately, the current rules of our marketplace give lumber companies incentives to liquidate their assets (cut down all their trees) when wood prices stagnate. Cutting into the growing stock of the forest is like damaging the goose with the golden eggs, diminishing the capacity of the system to provide year after year.

A sustainable society, therefore, depends not just on shifting to renewable resources but on carefully managing all of the Earth's regenerative gifts. This will require not only a taboo against exploiting ecological capital but also a reining in of the forces that drive this exploitation: runaway population and consumption. In short, it will require a simpler and more graceful life.

8. Remain in Balance with the Biosphere.

When we talk about a prairie or a redwood forest, we're talking about subcycles churning within a much larger cycle. The grandparent of all cycling occurs at the level of biosphere.

The biosphere (the layer of air, land, and water that supports life) is a closed system, meaning that no materials (except for rogue meteorites) are imported or exported. The stocks of the major biochemical building blocks such as carbon, nitrogen, sulfur, and phosphorus stay pretty much the same, even though they are actively traded among organisms. Whatever is removed from the resource reservoirs, through the process of photosynthesizing, respiring, growing, mineralizing, and decaying, is replaced in equal amounts. Through the revolving door of organisms, the stocks circulate but they don't run down.

Gases in the atmosphere are also held in a delicate but dynamic balance. In photosynthesis, plants inhale carbon dioxide and exhale oxygen. Respiring animals take this same oxygen and return carbon dioxide to the atmosphere. Neither of these gases is removed or returned in excess; for example, oxygen remains at a crucial 21 percent level in the atmosphere (which is a comfort to us every time we strike a match). A similar stabilizing effect is seen in the nitrogen, sulfur, and water cycles.

Through this give-and-take, life maintains the conditions needed for life. If these biological processes were to cease, writes environmental economist Robert U. Ayres, "The grand nutrient cycles would wind down, as the many chemical reactions proceeded toward chemical equilibrium." In short, the great juggling act of life would end.

Remaining in Balance with the Biosphere: The Lessons Learned

As living beings, we contribute our share of exhaled gases and organic matter to the Earth. Unfortunately, our byproducts go far beyond these bodily offerings. As Ayres reports, "In the middle of the last century about 280 of every million molecules in the atmosphere were carbon dioxide [CO_2] molecules (280 parts per million by volume). Today that value has risen by 25 percent, to about 355 parts per million by volume. The current rate of increase is about 0.4 percent per year."

Where does it all go? The 7,100 million metric tons of carbon per year that we inject into the air by fossil fuel burning and deforestation is only about 12 percent of the net primary productivity—the 60,000 million metric tons of carbon that land plants produce in their bodies each year. But while the carbon that plants produce is eventually reused by living things, our injection of CO_2 is not balanced by natural processes. Because it is over and above what would be naturally recycled, the CO_2 concentration in the atmosphere just keeps growing. The ultimate question that industrial ecologists must ask is: How on Earth will our biosphere respond to this perturbation in the grand nutrient cycle, this buildup without balance?

Industrial ecologists say the only answer is an industrial ecosystem that can dovetail with the biosphere without harm. A few minds are talking about this large-scale integration, but at this point, the talk is still talk. Ayers writes that unlike the Earth's system, which is characterized by closed cycles, the industrial system as a whole is an *open* one in which "nutrients" are transformed into "wastes" but are not yet significantly recycled. Like any linear system (such as the flour beetles in a bin), this one is inherently unstable and unsustainable. Without blinking, he writes, "It must either stabilize or collapse to a thermal equilibrium state in which all flows, that is, all physical or biological processes, cease."

Ayres cheers up somewhat when he reminds us that the Earth was not always a closed system. It took billions of years to evolve all the weblike mechanisms (organisms) that sew up the cycles. Before they were in place, the world faced lots of loose ends: organic molecule shortages (as protocells forming in the ocean used up all their building blocks), carbon dioxide buildups (before blue-green algae were around to breathe CO_2), and near oxygen poisoning (before aerobic bacteria were around to breathe O_2). So life has stood at the brink before, Ayres tells us. What will evolve to pull us back this time?

The solution that would save the day, he says, would be very difficult for us to predict. For that matter, so would "the last straw." The problem is that the biosphere and our industrial worldwide ecosystem are both complex systems, meaning that small changes can amplify to become very large changes. The most popular example of this "sensitivity to initial conditions" is the complexity of weather; theoretically, a butterfly flapping its wings in Central Park could trigger a series of disturbances leading to a typhoon in Taiwan. This nonlinearity makes it tough to gauge the seriousness of our current insults, or to predict the outcomes of our intervention.

All we can do is watch for warning signs. To that end, we are now monitoring the Earth more closely than ever before, hoping to discern patterns in how we affect the biosphere and how it responds. One of the largest new efforts is NASA's Mission to Planet Earth begun in 1991. (The mission was instigated by astronaut Sally Ride, who noticed that we spend millions to monitor other planets, but very little to track changes here at home.) In the mission's first phase, a number of new remote-sensing satellites are tracking, for example, circulation patterns of the world's oceans, weather disturbances caused by El Niño, fluctuating sea levels, shifting boundaries between temperate and boreal forest types, and the effects of CFCs on the ozone hole. Phase II begins in 1998, with the launch of the first Earth Observing Spacecraft, which, together with the satellites, will beam back more information every hour than currently exists in all the Earth sciences combined. If we use this information correctly, it could just be the self-regulator we've been searching for.

9. Run on Information.

Mature communities, like innovative and productive companies, have rich communication channels that carry feedback to all members, influencing their march toward sustainability. Excess and waste are held in check by mechanisms that reward efficient behavior and punish foolish genes. Any organism that is surrounded by and dependent on so many other links must develop unambiguous ways to signal its intentions and interact with its neighbors. Wolves, for example, must perfect ritualized gestures that clearly state things like "Let's mate" or "You win, I'm moving away peacefully." As biologists say, successful body designs and behaviors must be high in information content.

What makes a mature community run is not one universal message being broadcast from above, but numerous, even redundant, messages coming from the grass roots, dispersed throughout the community structure. A rich feedback system allows changes in one component of the community to reverberate through the whole, allowing for adaptation when the environment changes. The *raison d'être* of mature communities, remember, is to maintain their identity throughout environmental storms and travails, so they can remain, and evolve, in place. This is what sustainability seekers are also beginning to want for our communities.

Running on Information: The Lessons Learned

System breakdown occurs when we as a species ignore the negative feedback coming from the natural world—the reproductive abnormalities, the drastic weather changes, the extinctions—and ratchet up our growth gears anyway. We take more than the world can replace and release more than it can handle. This damn-the-evidence kind of excess is called "overshoot."

To avoid overshooting, all the firms in an economy have to be keyed into each other and aware of their interactions with the environment, the way organisms are. What we need to establish are feedback links among and within businesses, as well as feedback from the environment to business.

The recent proliferation of materials exchange brokerages such as the North East Industrial Waste Exchange in Syracuse, New York, and BARTER (Business Allied to Recycle Through Exchange and Reuse) in Minnesota is a good sign. These companies publish up-to-the-minute catalogs of who needs what and who has what, matching companies looking to get rid of wastes with those who could use those wastes. This recycling yenta service signals the beginning of information postings within and between industries that will facilitate the thorough use of materials. Such a feedback system could keep materials cycling through the economy instead of landing in a dump or incinerator.

Feedback within a firm can also help improve environmental report cards. Back in the fifties, breakthroughs in cybernetic feedback made automation possible. The heater in your home is a chip off that block—its information relay is a thermostat that senses the temperature in your house and turns the heater on or off so you don't have to. Industrial ecologists would like to see the same sorts of self-policing mechanisms built into machines to help industry avoid environmental transgressions. These mechanisms might constantly monitor emissions, for instance, calibrating the machines to be as clean as possible.

Feedback mechanisms need not be only mechanical, however. Profits falling or rising in response to a company's environmental record can also be a damping or driving mechanism. Governments can help push profits the right way by taxing firms for environmental transgressions and rewarding them for advances.

Another feedback mechanism that would concentrate industry's mind is consumer demand for greener products. The countries in the European Union, the United States, Canada, and Australia are currently negotiating an enforceable and credible green labeling scheme. Once a green seal of approval on a product becomes truly coveted, companies will go all out to make their products "greener than thine."

Finally, say the industrial ecologists, we need a response system that will allow firms to get warning signals from the environment and immediately clean up their practices without waiting for laws or profit to do the job. The government-industry covenants in The Netherlands are a good example of negotiations that are adaptive in nature. The Dutch decided that they wanted to reach their clean environment goal within a generation. They felt that the practice of legislating change involved too much guesswork and often didn't go far enough. With covenants, the firm tries out an environmentally friendly policy, and scientific monitoring shows whether it's working. If the environment is still suffering, there's no need to wait for new legislation to toughen policy—government and industry simply negotiate a speedy change in the covenant.

10. Shop Locally.

Because animals can't import products from Hong Kong, they shop locally and become local experts in their own backyards. Mountain lions coevolve with mountain goats, for instance, developing a "search image" for their prey and the perfect complement of physique and teeth needed to harvest and digest them. The goats, for their part, are equally adept on home turf, where they have evolved clever defenses to an enemy they know. Staying close to home is a coup then—it conserves energy and makes for the best use of an organism's abilities.

For that reason, say Allenby and Cooper, "Biological communities are, by and large, localized or relatively closely connected in time and space. Thus, for example, the nutrients in a rotting log are carried into the soil by rainwater, using energy from sunlight captured as the water initially evaporated. The energy flux is low, the distances proximate." In other words, with the exception of some high-flying migrant species, nature doesn't commute to work.

Shopping Locally: The Lessons Learned

Shopping locally is one lesson that we seem to be ignoring completely. The drive is now on for a global, borderless economy where a single product is assembled in a dozen different countries, and foods, even those that could be grown next door, are trucked, flown, and shipped from foreign soils (the average piece of food on your table was transported fourteen hundred miles). There are at least three problems with this approach. First, this way of life assumes that a transportation system, with its inherent energy greed, will always be available to us. It may not be. Second, having the whole globe in your backyard encourages regional populations to grow beyond what the land would allow if there were no imports. And third, when you separate producers from consumers, the consumers lose a visceral sense of where their resources come from, and what it costs environmentally to provide them. Deforestation in Third World countries is out of sight and mind, present only in books about the rain forest displayed on teak coffee tables.

If we were to take a page from nature's book, we would try to adapt our appetites to where we live, getting our resources from as close by as possible. Shopping locally requires a local knowledge that indigenous people possess but that many of us have lost. (A typical bioregional quiz asks: Do you know what watershed you are living in or what kind of vegetation your backyard used to support? Do you know what you would be eating if you were dependent on local larders?)

The good news is that local self-reliance movements are popping up like mushrooms. People are educating themselves about their natural addresses and trying to become, as environmental author Kirkpatrick Sales says, "dwellers in the land." Marketing co-ops are encouraging shoppers to buy locally produced foods, wood products, artwork, and literature as a way to sustain local economies. If this bioregionalism movement achieves its full promise, economic borders will be redrawn in real terms, being more closely related to watersheds, soil types, and climate regimes than to the political boundaries we currently honor.

The idea of an economy that suits the land and takes advantage of its local attributes would bring us closer to mirroring organisms that have evolved to be local experts. Instead of "becoming one patch," says William Cooper, we would assume the more stable patterning of an ecosystem—a mosaic of unique patches, each pulsing to its own rhythm, in sync with its own place.

Despite the commonsense rightness of the ideas in our list, leaping from Type I to Type III systems is at this point still a sport for "early adopters" or companies that can afford to experiment with new paths. How do we win over the critical mass of companies that will be needed for a full-web industrial ecosystem?

One way is to simply wait to hit the wall. As Cooper says, "Once we start to reach the Earth's real carrying capacity, today's market-driven ideas of maximizing flow-through will become institutional fossils real quick. We'll have to dump them for another way of doing business." Moving from a Type I to a Type III strategy, he says, will require wholesale substitutions, not slow developmental change. What will the changeover look like?

"It's called anarchy," Cooper says, half wisecracking, half serious.

Brad Allenby has a little more faith. If we can absorb the lessons found in nature, he thinks we can actually self-correct before we find ourselves, along with many other species, dangling over the brink. When I ask him what mechanism we can use to nudge our gigantic many-armed octopus of an economy toward sustainability, he smiles. Like many phenomena and patterns in nature, Allenby's approach is self-referential.

"We use the economy itself."

GETTING THERE:
SOME NICHE-SHIFTING TOOLS

Boundary Conditions

The industrial ecologists I talked to admit that we will always need some command-and-control laws like the kind that banned lead in gasoline or phased out CFCs. But that's not the whole answer. I wondered briefly if Allenby thought corporate America would simply volunteer to go from brown to green.

"We needn't ask industry for altruism," he replied, "and that's good, because it's not in their nature." If we structure things properly, Allenby believes, profit can be used both as cattle prod and carrot for the cultural and technological evolution we are seeking.

"A company that is profit-oriented has all kinds of handles and buttons—you know how to push and pull it—how to induce the kind of behavior you want." Market-based carrots and sticks wielded by governments, for instance, are one way to herd or nudge the system toward sustainability. Another is take-back laws and community right-to-know legislation. These regulations act as "boundary conditions." They place industry in a new operating environment, a new habitat, where environmental care is suddenly the most natural and competitive way to behave.

Cooper explains boundary conditions in biological terms. "If you put a species in an Illinois cornfield, it's going to act differently than if you put it in a beech-maple forest habitat. The conditions are different, the biological checks and balances are different, and natural selection will reward different behaviors in the survivors."

In the same way, if we were to plunk our economy into a set of conditions (incentives, disincentives, laws) that more accurately reflected real consequences and limits, it too would react and adapt. Right now, we've been smoothing the playing field artificially. We haven't been including costs to the Earth or to future generations in our accounting. Even worse, we've been subsidized negative activities. Fossil fuels have been subsidized worldwide to the tune of $220 billion annually. Artificially low prices give us a false notion of abundance and blind us to the true danger of depending on nonrenewables. What if we were to remove the rose-colored glasses and let the economy actually see the world in black and white? What if we were to resume play on a field that had all the ankle-twisting potholes of environmental limits?

Allenby thinks realistic boundary conditions would bring out sustainable behaviors in the economy, just as boundary conditions associated with a mature forest—a finite amount of water, sunlight, and nutrients—bring out stability-inducing characteristics in its members. In effect, it would give Adam Smith's invisible hand of capitalism a green thumb.

Allenby believes in the power of boundary conditions because he knows how futile it is to micromanage specific elements of the system. "We learned in the command-and-control era of the seventies that the system is just too complex for us to know where to intervene effectively." Instead, he thinks government should draw the lines—the minimums and maximums that society will allow—and invite industry to color within the lines in whatever way it sees fit.

In Allenby's model, laws that mandate a certain kind of compliance technology would be struck from the books, giving companies the freedom to explore and come up with even better solutions. Antiquated subsidies that reward excessive deforestation and mining would also have to be removed. In their place, Allenby suggests "very broad, nonprescriptive policy tools that push the industrial system in the desired direction, without trying to define the endpoint, either organizationally or technologically." Instead of a detailed map, he says, "we should draw an arrow and dare companies to get there before their competitors." Now, that's in their nature.

Boundary conditions are a great start, but if the system oscillating inside those boundaries is to land softly anytime soon, it has to have all its internal signals blinking clear and true. In our economy, that means having products' prices truly reflect costs to the Earth and to future generations.

Green accounting would have strong and immediate effects. "Think what would happen to agriculture if the price of water floated up toward its real social and environmental cost," says Allenby. "You'd pay a king's ransom to grow a thirsty plant like cotton in the San Joaquin Valley! Instead, farmers would most likely switch to crops that were better suited to the region." And what if industries were forced to cover the full environmental costs of their activities up front instead of leaving the bill for the public to pay? It would no longer be a question of whether environmental transgressions were really serious—they would be expensive, and therefore enough of a thorn to warrant pulling. On the other hand, environmentally benign technologies would be coveted, because in this new schema of pricing, green manufacturing would actually be cheaper.

Government, in its role as tax collector, might play a natural role in rewiring our economic steering wheel to its drive train. Paul Hawken thinks we've had it backward. Instead of taxing *good* things like income, Hawken would like to see government tax *bad* things like pollution or excessive use of energy or virgin materials. Taxing fuels based on their carbon content, for instance (more carbon is more damaging), would encourage use of low-polluting fuels like natural gas in every stage of a product's life cycle. The price on nonrenewable raw materials would be raised to more realistic levels, discouraging waste and giving incentives for recycling. The positive flip side would be to give tax credits to companies that are producing renewable resources in a sustainable manner.

Government can also reward early adopters through its own purchasing practices. The Clinton administration swung a huge green club when it required the federal government to give preference to green, recycled, energy-efficient products in its procurements. When a customer the size of Uncle Sam goes green, the makers of computers, office supplies, vehicles, and more suddenly rush to come out with a qualifying product line.

Another governmental program that has shown faith in the invisible hand is the scheme, introduced in the 1990 Clean Air Act, to create markets in tradable "right to pollute" credits. Here's how it works: The government issues a limited number of pollution credits to companies—saying, that's as much as you can emit. Companies that figure out how to cut emissions no longer need their credits, and can sell them at a Chicago Board of Trade auction (the first one was held in 1993), collecting payment from companies that haven't been as innovative. Suddenly, bad environmental practices are no longer just costly; now they force you to (ouch) line the pockets of your competitor.

Once a critical mass of companies begins to clean up their environmental messes, we could see change beget change, as in a positive feedback or snowballing effect. The companies that cut their emissions, for instance may suddenly become "reformed smokers," advocating stronger laws that would force other companies to scramble to keep up. Responding to pressure from above, within, and below, the swirling mass of our economy may begin to realign into a Type III community bent on optimizing rather than maximizing.

One of the ways we could speed this transition is to make sure all our signals are blinking unambiguously that "being green is good for business." Job one is to change the way we measure our economic well-being. Right now, we genuflect to the GNP, which is not a measure of health so much as it is a measure of exchange. It tracks flow-through of materials, and it rings positive when we are using up resources as fast as we can. Even negatives, like pollution, cancer, and other ills, are seen as positives so long as we keep cranking out products to deal with the cleanup or the cure. In this system, the *Exxon Valdez* goes aground, and the GNP jumps (true story).

There is, thank goodness, a movement to find a new way of monitoring economic welfare, and it's called Green (like everything else in this movement) GNP. As a first step, the U.S. Department of Commerce's Bureau of Economic Statistics is inventing a way to put a dollar value on environmental assets, a new column on its output and investment ledgers. Other countries are also experimenting with report cards that might take into account a wide range of social, economic, environmental, and health-related factors, such as life expectancy, infant mortality, the general health of the population, literacy, crime, accumulated wealth, income distribution, air quality, water quality, and recreational opportunities.

Meanwhile, at the level of individual firms, Allenby thinks environmental costs that were once buried in overhead must become a part of every department's debit and asset sheet. People at the drawing board, for instance, will have to know what their design choices will cost in terms of the environment. An engineer who orders a cadmium-coated fastener will have to consider more than just price and function; after factoring in the environmental headaches of working with a hazardous compound, he or she may decide that a noncoated fastener, even if it's more expensive, may be worth the cost.

If only we had known more about environmental costs sixty years ago when chlorofluorocarbons (CFCs) were invented. Environmental advocate Hazel Henderson estimates that the true societal cost of *one* CFC aerosol can—factoring in its contribution to the destruction of the ozone layer, cancer rates, and so on—would be about $12,000. That might have given its designers pause.

Making Envy Green

When you think about it, designing may be the most powerful fulcrum from which we can move the economy and the culture toward a more sustainable place. Designers are the people who give a product not only its functionality but also its personality. From art-deco lamps to the tail fins on Cadillacs to Euro-style Bang and Olufsen stereos, designers have been trained to capture the dreams and aspirations of society—what we are or hope to be.

Also embodied in their designs is the record of our relationship with the Earth. For the most part, the disposable and energy-hungry products that litter our homes loudly trumpet our disregard for other living things. What if designers could help us alleviate some of this psychic guilt?

Christopher Ryan, professor of design and environmental studies at the Royal Melbourne Institute of Technology in Australia, thinks that designs that are deeply green—in manufacturing, use, and "afterlife"—will give people the option of enjoying life without destroying it, getting the service they want without the bloodbath of consequences. Once a few choice designs show people this guilt-free alternative, says Ryan, passivity is no longer acceptable. Just as safety considerations are now an expected part of any design, people will want to know why greenness can't be incorporated into all products.

Designers, together with marketing experts, can help make green de rigueur by first making it look more fun. Ryan feels designers lost their chance in the seventies, when environmental friendliness was packaged to be about as exciting as wearing a hairshirt on a summer's day. Today Ryan thinks we have a new chance to make envy green—to make environmentally friendly products so fashionable that everyone will want a green product. In this way, the design of sustainable products may actually precede the sustainability revolution and help bring it into being.

Design for the Environment

Successful design also has to pass another test beyond consumer appeal. It has to enhance the company's bottom line on the manufacturing side as well. That's why Allenby and Graedel are working on Design for the Environment tools to help engineers and operating managers build environmental friendliness into every step of the production process, as well as into the product itself. The first is a matrix-based approach that allows a manager to assign a green score to a potential product or process. The completed matrix is a spreadsheet full of ovals, some filled in and some empty, like the ovals of a *Consumer Reports* score card. The darkest ovals tell process and product engineers where the greatest areas of environmental concern are, and, subsequently, where the greatest green strides can be made. It brings the impact on the environment into the equation, giving engineers a green set of sieves through which to filter their ideas.

Graedel has also invented a version of Life Cycle Analysis (LCA) that allows engineers to pit two products against each other, using actual numbers instead of just ovals showing relative levels of concern. His LCA would calculate, for instance, the kilowatts of energy used at each and every stage of a product's development, from scooping the oil from the ground to the cost of reincarnating it after its use. This cradle-to-cradle accounting is great for comparing two products, such as cloth diapers vs. disposable ones (it's still a toss-up). The advantage of Graedel's analysis is that it can be done in a couple of days rather than a couple of years, which is how long most LCAs now take. The only problem with the new tool may be the stampede of industries that want to try it.

BUSINESS CAN BE A JUNGLE: THE PROMISE OF INDUSTRIAL ECOLOGY

LCA-inventor Tom Graedel is a soft-spoken man whose eyes are lit with a keen intelligence tempered to a patient, steady flame. He makes an unlikely maverick. "I'm actually a little afraid," admitted Graedel when I asked him about the tsunami of interest that had built around his life-cycle analysis tools. "Now that our textbook is out [the first of its kind in industrial ecology], we're going to be deluged." Checking back with him a few years later, I find he is indeed busy. Virtually every industrial field wants to know more about how it can design in greenness at the beginning to avoid corporate mistakes and win the badge of corporate environmentalism. Within AT&T, Graedel is making the rounds, visiting one department after another with his shortcut LCA tool.

Despite the demands, Graedel is happy to be on the ground floor of industrial ecology. His background is in atmospheric science, specifically atmospheric chemistry and climate change. Widely recognized as an expert, Graedel was consulting scientist in the project to recondition the Statue of Liberty after years of corrosion in New York City's air. He told me that when he stood at the feet of the statue for the unveiling ceremony, he thought that he was experiencing the absolute peak of his career. Now he thinks differently.

"When I look back, I think helping industrial ecology get off the ground will be by far the most important thing I've ever done. Industrial ecology has the potential to remake industry and, in our tonier thoughts, to remake society as well."

As I listen, I'm reminded of Laudise's comment to his peers that "industrial ecology has the capacity to change not only the way we make things but the way the world works." And later, to me alone, "Industrial ecology has a great deal to offer and I would love to see people appreciate and understand it at a poetic level if nothing else." Neither of these men strikes me as a dewy-eyed idealist. What they see I am beginning to see also—that our economy is fertile ground for making the inside-out changes that need to occur if we are going to mesh gears with the Earth and manage a soft landing.

Though it seems worlds away from ecology, industry might just be the perfect place to start pulling our rip cord. As Christopher Ryan writes in an Internet post titled *Green Goods*, "As we manipulate the materials we extract from our environment, through our industrious efforts, we engage in our most fundamental relationship with nature, that of its reconstruction. Every material thing we create, everything we produce, reflects our relationship to the physical and biological world." Right now that relationship is estranged and characterized by abuse. Remaking it into something that will sustain both the human race and the Earth's integrity is the great hope and the true mission of industrial ecology.

Our desire is greening. What we choose next can either satisfy our urge to do right by the Earth or plunge us deeper into denial. The biomimics, by holding up living, breathing examples of sustainability and daring us to emulate them, have become beacon-bearers at a crucial juncture, lighting the runway home.

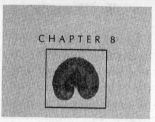

CHAPTER 8

WHERE WILL WE GO FROM HERE?

MAY WONDERS NEVER CEASE: TOWARD A BIOMIMETIC FUTURE

> *Humanity needs a vision of an expanding and unending future. This spiritual craving cannot be satisfied by the colonization of space. . . . The true frontier for humanity is life on Earth, its exploration and the transport of knowledge about it into science, art and practical affairs.*
>
> —E. O. WILSON, author of *Biophilia and the Conservation Ethic*

> *To sit with the eagles and their flutelike songs, listening to the longer flute of wind sweep through the lush grasslands, is to begin to know the natural laws that exist apart from our own written ones.*
>
> —LINDA HOGAN, author of *Dwellings*

As I put the finishing touches on this book, two households of geese fuss in the pond right outside my window. They are restless lately with what biologists call *Zugunruhe*, which means "travel urge."

Eleven goslings were raised this year, which is eleven more than last year or the year before. When I bought the property, everyone told me that the pond was a legendary nursery for waterfowl—cinnamon teal, blue-winged teal, mergansers, coots, and Canada geese. Two years ago, as you'll remember from Chapter 3, the once sparkling water was eclipsed by a solid sheet of duckweed, a tiny floating plant that forms colonies and manages to shade out everything below it.

It seems that duckweed in profusion is too much of a good thing, and birds that would normally relish it wouldn't even land on it. For two years, pair after pair wheeled over the pond at nesting time but opted to go elsewhere. I tried to remedy things by screening off the duckweed with a series of handmade contraptions, but like the Sorcerer's apprentice, I managed only to create more duckweed.

The county extension agents recommended that I treat with chemicals, but I had seen too many turtles periscope up, leaves like sequins on their sleepy lids, to even consider it. When I asked the agents for a more natural way to rejuvenate the pond, they were stumped.

Finally this summer, after one heaping wheelbarrow too many, I simply stopped. I stopped trying to engineer schemes in my own mind and I just sat down on the banks. I indulged in a fantasy of how I would like the pond to be—clear, loud with the squabbles of nesting birds, a healthy balance of vegetation and open water.

It was then that I became a biomimic instead of just writing about them. I realized that it wasn't a fantasy I was referencing, but an actual place, a pond that I had once biked to up near the National Forest. I peeled off my swamp boots and got on my bike.

I spent the afternoon on the lush banks of that balanced pond, trying to absorb its secrets. I noticed the way grasses and willows crowded at the edges, and, when I dipped my hand in, I found it sharply colder than my pond. My final clue came when a cottonwood leaf cruised lazily into view—and out again. Current!

The only times I could remember seeing current in my pond were during the spring chinooks when snow melts in a hurry and brings muddy waters from the surrounding fields. A few times each year, these floods would turn the pond a Mississippi brown.

By then it was coming clear to me. My pond must have originally been spring-fed, but lately the source of fresh water, the maker of current and cold, had been suffocated under layers of topsoil roiling in from the fields. The topsoil was prime for eroding because years of overgrazing had weakened the thick sod. One thing led to another, and the pond silted in, becoming a tepid bowl—perfect for duckweed but not, ironically, for ducks. If I wanted to keep the pond open to breeders and have duckweed only in the cattailed edges again, I would have to find that forgotten spring, free it, and then stop the source of silting.

I went home and gave my neighbors one more thing to talk about as I slowly paddled through the green froth, feeling for the coldest spot. I started to dredge there, and sure enough, great shovelfuls of good topsoil came up. What came up next felt like a miracle.

Released from its burden, a cleansing swell of cold Montana snowmelt geysered to the surface. The once-murky waters rose to fill their banks, and the duckweed I had labored to screen away for two years flowed casually over the dam in sheets. By afternoon, my pond was sparkling, and the wood ducks in the sloughs of the river below me were feasting.

Mine was a classic example of echoing nature, and if I were to offer some sort of path for the larger culture to take toward a bio-mimetic future, it would have to follow this pattern. Like all echoing, mine was a dialogue with the land, but instead of me speaking and a canyon amphitheater responding, it was the other way around. I listened while the land spoke, and then I tried to mimic what I had heard.

The preparation for this echoing was a *quieting* on my part, a silencing of my own cleverness long enough to turn to nature for advice. My afternoon vigil at the pond was the *listening* stage, the absorbing of secrets in a respectful way. My uncovering of the forgotten spring was the *echoing*, the biomimicking itself. The follow-through to all this was the *stewarding* required of me, an ongoing thank you for the wisdom I had acquired. It was up to me to reve-getate my denuded lands with native plants that would hold the soil so that flood events would not continue to suffocate the spring.

In my adventure with the pond, I realized that biomimicry is just like opening a forgotten spring, rushing new hope to problems that have seemed intractable. The steps to the biomimetic future that I offer below are based on that experience. They are part stu-dentship and part stewardship—studying nature's wellsprings of good ideas, and then protecting them so that they can continue to flow.

FOUR STEPS TO A BIOMIMETIC FUTURE

Quieting: Immerse ourselves in nature.

A solitary American monk named Thomas Berry writes that in our relationship with nature, we have been autistic for centuries. Wrapped tightly in our own version of knowledge, we have been unreceptive to the wisdom of the natural world. To tune in again, to have the "spontaneous environmental rapport" that characterized our ancestors, will take doing something that is perfectly delightful: reimmersing ourselves in the natural world.

Our first taste of nature usually comes in childhood, but even that must no longer be taken for granted. Sadly, report Gary Paul Nabhan and Stephen Trimble in their book *The Geography of Childhood*, it is quite possible for today's child to grow up without ever having taken a solitary walk beside a stream or spent the hours we used to "doing nothing," foraging for pine cones, leaves, feathers, and rocks—treasures more precious than store-bought ones. Today it is difficult to tear children away from the virtual world of the mall to introduce them to the real one.

Bringing children back into nature and nature back into childhood is a job for teachers and parents and friends willing to take a child outside for a lark. There need not be an "official" park involved; finding a place where green things grow, even if it's a crack in the sidewalk, is enough. Once you're there, there's no need to "do" anything. What kids really need are large blocks of unstructured time for making mud pies and finding nests, for acting on the fascination with nature that is part of our reptilian mind, and that is thankfully still unsullied in children.

As adults, we need to put down our books about nature and actually get into a rainstorm, be startled by the deer we startle, climb a tree like a chameleon. It's good for the soul to go where humans do not have a great say about what happens. Between these trips to the "big outside," we need only open our hearts to the smaller encounters: the smell of old sunlight in a leaf pile, the chrysalis of a butterfly inside our mailbox, the glimpse of that earthworm that helps us grow tomatoes.

This literal immersion in nature prepares us for the figurative immersion. This is where we take our reasoned minds and stuff them back into our bodies, realizing that there is no membrane separating us from the natural world.

For a long time we thought we were better than the living world, and now some of us tend to think that we are worse, that everything we touch turns to soot. But neither perspective is healthy. We have to remember how it feels to have *equal* standing in the world, to be "between the mountain and the ant . . . part and parcel of creation," as the Iroquois traditionalist Oren Lyons says.

We may be newcomers here, but we are not aliens. The old-time Montanans have taught me that when dealing with new residents, the important question is not "When did you get here?" but "How long are you planning to stay?" If we are planning to stay forever, we need to look to the life that has preceded us for tips on how to be better neighbors.

Listening: Interview the flora and fauna of our own planet.

I say "interview" because it is not enough to simply name the species on Earth (though this in itself is a monumental task, and nowhere near complete). We must also get to know these species as best we can and discover their talents and survival tips, their role in the great web of things.

Entering into this sort of intimacy with life on Earth is not a job for scientists alone. It calls for a renewed popular interest in natural history, like the flourishing of nature-buffism that characterized the 1800s. In those days, amateur naturalists contributed mightily to the literature, and nature study with hand lens and plant press was a common form of family recreation. I see some hint of this in people's increasing desire to get to know their own region better, to develop an internal pride of place. Naturalists tell me that people are showing up at mushroom walks and owl-calling nights and master-gardening lectures in droves, finally curious about the true nature of their homeplace.

At the same time, the pool of people who can teach these courses is thinning out. In one of the most frightening essays I've read in a long time ("Forgetting"), Rutgers University's David Ehrenfeld talks about how many basic courses such as Classification of Higher Plants, Marine Invertebrates, Ornithology (birds), Mammology, Cryptogams (ferns and mosses), and Entomology (insects) are going untaught at highly respected universities. They are no longer offered because there is no one on the staff qualified to teach them. Retired faculty sometimes volunteer in a pinch, but few graduate students are following in their footsteps. For many students, systematics is not a glamorous enough career track, nor well enough funded. Ehrenfeld asks administrators to realign their priorities, to make sure that before they break ground for one more "world class molecular biology lab," they figure out how the fundamental knowledge of our natural world will be passed on.

Harvard's E. O. Wilson has the same concerns, especially as he contemplates the great exploration that still lies ahead. In an April 1989 *Bioscience* article, he writes, "Systematics [deep knowledge of particular groups of organisms] in the broad sense is the future of biology. The responsible expert is the steward of a chosen taxonomic group. . . . He or she knows best which organisms exist and where, which are most endangered, which offer new kinds of problems to be solved, and which are most likely to benefit humankind. No one but the systematist can reveal the particular and extraordinary value of alcyonacean corals, chytrid fungi, anthribid weevils, sclerogibbid wasps, melostomes, ricinuleids, elephant fish, and so on down the long and enchanted roster."

With at least 30 million possible names on that roster, we could use a peacetime army of people trained in basic identification and observation skills. As long as we're enlisting, I would love to see a Biological Peace Corps that would give adults of all ages a chance to volunteer for two years in this very important work.

Our interview with life will allow us to become "innovation matchmakers," matching nature's designs and processes to the needs of technologists and engineers who design the shape, feel, and flow of our products, materials, and systems. We are at a turning point right now in which much of this century's infrastructure is in need of replacement, including outmoded highways, energy and communication networks, water treatment facilities, factories—even economic models. This time when we collect proposals for public works and policies, we want to make sure that nature's blueprints are at the top of the stack.

Echoing: Encourage biologists and engineers
to collaborate, using nature as model and measure.

The only way to ensure that nature's designs will be considered is to put biologists and engineers on the same working teams. Unfortunately, many engineers I know say they're not interested in the life sciences, just as many of my biologist friends profess boredom at all things mechanical. I find that odd, because as the biomimics in this book taught me, life manufactures, computes, does chemistry, builds structures, designs systems, and engineers, to within a fine tolerance, the tools needed to fly, burrow, build dams, heat or cool their homes, and so on. The difference between what life needs to do and what we need to do is another one of those boundaries that doesn't exist. Beyond matters of scale, the differences dissolve.

The trick is to show this hidden likeness to engineers and biologists before they put on their blinders. It takes educating in the estuary—the place where two or more disciplines flow together to make a fertile ideabed. Throughout their degree work and even in their continuing education, biologists and technologists should take courses in one another's fields. At think tanks, task forces, joint forums, conferences, and professional societies, they should get to know one another on a personal basis, rubbing minds and getting a little creative friction going. Sparks fly from these mixed unions in a way that just doesn't happen within bureaucracies of like-minded people.

To encourage interaction on an ongoing basis, universities would be wise to create interdisciplinary departments for the express purpose of making the metaphors flow the right way, from biology to engineering.

Until that happens, there are many ways we can make biological knowledge available to innovators wherever they are. Using the Internet, for instance, systematists could maintain a giant database of information about known taxonomic groups—their biochemical makeup, their ability to survive in certain conditions, their flight speed, and so on. Engineers could work with the biologists to help design the categories of information in the database to make sure that searches would be useful to them. This way, an engineer charged with designing a new desalination device, for instance, could easily review the strategies of mangroves and other plants that filter seawater with their roots.

Finally, when the biologist/engineer collaboration has yielded a new device or process or system, we should use what we are learning about nature's survival principles to screen for viability—meaning, quite literally, to judge whether or not our new solutions promote life.

For too long, we have judged our innovations by whether they are good for us, which has increasingly come to mean whether they are profitable. Now that we realize, as my Jamaican friend says, that "All of We is One," we have to put what is good for life first, and trust that it will also be good for us. The new questions should be "Will it fit in?," "Will it last?," and "Is there a precedent for this in nature?" If so, the answers to the following questions will be yes:

Does it run on sunlight?
Does it use only the energy it needs?
Does it fit form to function?
Does it recycle everything?

Does it reward cooperation?
Does it bank on diversity?
Does it utilize local expertise?
Does it curb excess from within?
Does it tap the power of limits?
Is it beautiful?

Assuming our bio-inspired innovation passes those tests, our next design decision will have to do with scale. Since scale is one of the main things that separates our technologies from nature's, it's important to consider what is appropriate, that is, what is receptive to and acceptive of our habitat. Wendell Berry's test of scale is simple but valuable. In his book of essays titled *Home Economics*, he writes: "The difference [between improper and proper scale] is *suggested* by the difference between amplified and unamplified music in the countryside, or the difference between the sound of a motorboat and the sound of oarlocks. A proper human sound, we may say, is one that allows other sounds to be heard. A properly scaled human economy or technology allows a diversity of other creatures to thrive." I find this last point compelling because any bio-inspired technology that diminishes diversity diminishes the very inspiration upon which it depends. By letting the diversity of life on Earth erode, we smother the wellspring of good ideas.

Stewarding: Preserve life's diversity and genius.

The erosion of life in this country alone includes: Ninety-five percent of all virgin forest cut down in the last two hundred years. Nearly all the prairie turned "wrong side up." Sixty percent of all wetlands drained and filled. And now, according to the new National Biological Survey, half of all native ecosystems degraded to the point of endangerment. It's no secret that we can level entire habitats as if we were sweeping children's blocks from a table. But can we refrain from doing that?

Restraint is not a popular notion in a society addicted to "growing" the economy, but it is one of the most powerful practices we can adopt at this point in history. Over the next several decades, we will be doubling our population before we begin to level off at 10 billion by mid-century. If most of us hope to live above the poverty level, our economy will have to somehow mushroom tenfold. This means more pressure on wild lands than ever before. If current rates of deforestation continue, for instance, only 10 percent of the original tropical forest cover will be left by mid-century along with only 50 percent of its biodiversity. The alternative, say biologists, is to plan now to save habitat and get wild species through the knothole of the next several decades.

This calls for a new valuing of what we have left, not in an economic way but in a much deeper way in which we acknowledge that we are ultimately dependent on the existing natural pattern, *a pattern that we only partially understand.* Science is continually peeling masks away, only to find another mask deeper down, one of the many worn by what Thomas Hardy called the Great Face behind. The closer we come to glimpsing that face, the greater the mystery appears. Our partial knowledge—the fact that we are, as Wes Jackson says, more ignorant than knowledgeable—is the best reason to save wild lands in their unadulterated state.

Native peoples' response to this abiding mystery was to set aside sacred sites—a valley that would not be hunted, a stream that would not be fished, a grove of trees that would never be cut. These spiritual sites turned out to be lasting conservation legacies. In some parts of the world, they are the only examples of a certain kind of habitat that are left.

But saving fragments of land here and there is not enough. One of the latest revelations from conservation biology shows us that an ecological unraveling occurs as the fabric of the landscape is cut into smaller and smaller pieces. The smaller the "island" of land, the more edges there are to fray, and the more vulnerable species are to human influence, genetic inbreeding, catastrophic disease, and ultimate extinction. One way to alleviate the isolation is to connect large blocks of wilderness with protected migration corridors. Blocks-and-corridors alternatives such as The Northern Rockies Ecosystem Protection Act are, in my opinion, the only land-use plans before us that honor ecological realities.

Important as they are, wild set-asides cannot preserve the lion's share of biodiversity, which resides in and among us in our settled lands—our urban forests, our suburban greenspaces, our farms and ranches. We cannot escape the fact that we must use nature on the lands; our lives depend on the lives of other species. As the author Wendell Berry says, the question will be how to use it and how to use. Again, our actions must be guided by a humility that from the realization of how little we know. There are four to five thousand species of bacteria in a pinch of ordinary soil—most of them species we don't yet have a name for, much less an understanding of why we need them. There must be studentship on settled lands, and only then, stewardship or good use.

The idea of good use also applies to how we use the *products* of those lands. Berry argues, for instance, that shoddy workmanship is a much greater threat to our forests than clear-cutting. Only when we come to value the well-made chair or table that lasts a lifetime will we begin to value and save the source of those things, whole forests instead of trees. When the product of that forest is a durable idea, as it is in biomimicry, the same valuing of source will occur.

Cultures that depend directly on hunting, gathering, and fishing tend to work out codes of behavior that honor both product and source. Richard Nelson, an ethnographer who has lived and hunted with native Alaskans, says there are literally hundreds of rules and rituals that keep hunters in the good graces of the animals they depend on.

Koyukon hunters believe that an animal either gives or withholds itself; success has nothing to do with the hunter's prowess. In fact, when a hunter returns to the village with a bear, he makes a cryptic comment like "I found something in a hole," so there is no semblance of boasting over the animal's demise. He then adheres to a strict ritual when butchering the body, beginning with the slitting of the eyes so that the powerful spirit of the bear will not see if the hunter makes mistakes. It is considered strictly taboo to deviate from this ritual, to kill more than what you need, or to waste any part of an animal.

Other rules maintain a sustainable harvest. Seine nets are sewn purposely wide to let smaller whitefish through. Beaver traps are designed to catch only large animals, and only two per beaver house are taken. These conservation ethics are based partly on ecological knowledge, says Nelson, and partly on the Koyukons' belief that the ⸴h is aware. As one elder told Nelson, "The country knows. If ⸴ong things to it, the whole country knows. It feels what's ⸴ I guess everything is connected together somehow, ⸴" Our science is also showing us, in a different way ⸴onnected country knows. The Gaia hypothesis ⸴ulates its own cycles and creates the conditions ⸴view, every part of the living world provides ⸴ing on one part of the web reverberates

To care for this knowing country is the ultimate act of gratitude, and it will be the sign of our maturity as a species. In her book of essays, *Dwellings*, Chickasaw author Linda Hogan writes, "Caretaking is the utmost spiritual and physical responsibility of our time, and perhaps that stewardship is finally our place in the web of life, our work, the solution to the mystery of what we are."

A SPECIES SHAPED TO ECHO

After meeting so many elegant beings produced by evolution, we ask at last what is noteworthy about *us* as a species. How do we contribute to the continuation of life? By virtue of asking the question, we partly answer it.

We whom Thomas Berry calls "the universe become conscious of itself" are self-reflective, and therefore uniquely positioned to seek nature's counsel, to learn, to echo, and to give thanks for the wisdom we acquire.

These self-reflective brains are evolution's latest attempt to find a way to handle and profit from information. At first, single cells floated in the broth, with information strung on the bonds between molecules. Next, a genetic code was devised to handle more complex information. As the gene pool evolved, organisms developed keener and keener senses to collect a stream of signals from the natural world. Finally, our brains became a powerful accomplice to those senses, allowing us not only to take in information (though scopes and satellites now) but to make it into Story—to see connections, patterns, and consequences and, finally, to envision a different future. This ability to mentally scout the river of time gives us an option: Run the rapids the way we always have, or pull into an eddy and learn a better way.

We're in luck here as well because learning is the second thing that we as a species are good at. Both as individuals strengthening our neuronal nets and as a society, accumulating an organic memory through science, art, and culture, we build on what we've learned. And, we have the capacity to seek this knowledge selectively, choosing who or what will teach us. If we choose to heed the lessons of the natural world, we become biomimics.

Which brings us to our third evolutionary gift. Like a larynx that is perfectly curved to boomerang a voice, we as a species are well shaped to mimic what we see and hear. Children guage and gender roles and acceptable behavior by mimicking adults, and as mimes they prove to be uncanny. The first artists were also practicing mimics, re-presenting the natural world in painting, song, and dance. The art of survival itself has probably always hinged on an ability to imitate the traits of the best and brightest in any habitat in which we found ourselves. Ice ages ago, hunters doused themselves in musk to smell like their prey, and today, native Alaskans still stalk seals by stretching flat on the ice like their polar bear mentors.

We are not the only species to have prospered through imitation; biomimicry has a long and colorful tradition in the living world. There are behavioral mimics like the cowbird chick, coloration mimics like the viceroy butterfly that resembles the poisonous monarch, and shape and texture mimics like the walkingstick, an insect that looks like a twig. Biomimicry helps animals and plants blend into their surroundings, or, in the case of the viceroy and monarch, to take on the traits of a species that is better adapted to its environment. By mimicking nature's best and brightest, we, too, have a chance to blend in and become more like what we admire.

In pursuing this path, we do more than ensure our own survival. In a world as interconnected as ours, protection of self and protection of the planet are indistinguishable, which is why the deep ecologists say, "The world is my body." If we act on our ability to mimic life's genius, we have a chance to protect both world and body. If we succeed, evolution will not have produced this giant brain in vain.

We are already off to amazingly good start, with so many instances of biomimicking cropping up that I couldn't include them all in this book: the proliferation of "green" communities built on ecological principles, the several hundred towns that have decided to use natural marshes to clean their wastewater, the restoration of Sacramento–San Joaquin River Delta and the Everglades by ⸱ ing natural flood cycles, the restoration of prairies and forests ⸱ wildfire and natural culling, and even a new political ⸱ the precept of using nature's laws to inform our own. ⸱ conscious emulation is occurring, drawing on the ⸱ still growing, knowledge of the natural world. ⸱'s know-how, we are reaching back to some very ⸱ n "urge to affiliate with life" that is embossed ⸱ ilson says that it's only natural for us to be ⸱ gs of the natural world. For the 99 percent ⸱ n Earth, we were hunter and gatherers, our lives dependent on knowing the fine, small details of our world. Deep inside, we still have a longing to be reconnected with the nature that shaped our imagination, our language, our song and dance, our sense of the divine. "To explore and affiliate with life is a deep and complicated process in mental development . . . our spirit is woven from it, hope rises on its currents," writes Wilson. He and others hope that this Biophilia, this love of life, will ultimately convince us to pull over and learn a better way.

In the end what makes us different from other creatures (as far as we know) is our ability to *collectively* act on our understanding. We can decide as a culture to listen to life, to echo what we hear, to *not* be a cancer. Having this will and the inventive brain to back it up, we can make the conscious choice to follow nature's lead in living our lives.

The good news is that we'll have plenty of help; we are surrounded by geniuses. They are everywhere with us, breathing the same air, drinking the same round river of water, moving on limbs built from the same blood and bone. Learning from them will take only stillness on our part, a quieting of the voices of our own cleverness. Into this quiet will come a cacophony of earthly sounds, a symphony of good sense.

The geese that were born here are honking their good-byes now, rising in a noisy ribbon that quilts the clouds with Vs. Deep in their genes is a map of mountains, sage steppes, grasslands, and riverbeds rolling like signposts along the curvature of the Earth. I follow the flock with my eyes until they are out of sight, clearing a ten-thousand-foot range with strong and liquid wingbeats.

In the silence their passage leaves, I begin to imagine that their farewell song was a kind of prayer, similar to the Mohawk blessing spoken by a midwife at the moment of birth: "Thank you, Earth. You know the way." Although the scientists and innovators I met might be hesitant to phrase it this way, it could just as easily be their journeying song. Together, we biomimics are setting out on a voyage to learn what nature's "long and enchanted roster" already knows. It's the way home, and I'm as eager as the geese to go.

BIO-INSPIRED READINGS

JUST THE TIP OF THE ICEBERG . . .

Berry, Wendell. *The Unsettling of America*. San Francisco: North Point Press, 1977.

Birge, Robert R. (ed). *Molecular and Biomolecular Electronics: Symposium Sponsored by the Division of Biochemical Technology of the American Chemical Society at the Fourth Chemical Congress of North America. New York, August 25–30, 1991*. Washington, D.C.: American Chemical Society, 1994.

Capra, Fritjof. *The Turning Point: Science, Society, and the Rising Culture*. New York: Bantam Books, 1982.

Center for Resource Management and David Wann. Introduction by Paul Hawken. *Deep Design: Pathways to a Livable Future*. Washington, D.C.: Island Press, 1996.

Chiras, Daniel D. *Lessons from Nature: Learning to Live Sustainably on the Earth*. Washington, D.C.: Island Press, 1992.

Etkin, Nina L. (ed). *Eating on the Wild Side: The Pharmacologic, Ecologic, and Social Implications of Using Noncultigens*. Tucson: University of Arizona Press, 1994.

Graedel, T. E., and B. R. Allenby. *Industrial Ecology*. New York: Prentice Hall, 1995.

Gratzel, Michael (ed.). *Energy Resources Through Photochemistry and Catalysis*. New York: Academic Press, 1983.

Gust, Devens, and Thomas Moore. "Mimicking Photosynthesis." *Science*, April 7, 1989, pp. 35–41.

Hameroff, Stuart R. *Ultimate Computing: Biomolecular Consciousness and Nanotechnology*. New York: North-Holland, 1987.

Hawken, Paul. *The Ecology of Commerce: A Declaration of Sustainability*. New York: HarperCollins, 1993.

Jackson, Wes. *Altars of Unhewn Stone: Science and the Earth*. San Francisco: North Point Press, 1987.

Johns, Timothy. *With Bitter Herbs They Shall Eat It: Chemical Ecology and the Origins of Human Diet and Medicine*. Tucson: University of Arizona Press, 1990.

Kellert, Stephen R., and Edward O. Wilson. *The Biophilia Hypothesis*. Washington, D.C.: Island Press, 1993.

Kelly, Kevin. *Out of Control: The New Biology of Machines, Social Systems, and the Economic World*. Reading, Mass.: Addison-Wesley, 1994.

Ogden, Joan M., and Robert H. Williams. *Solar Hydrogen: Moving Beyond Fossil Fuels*. Washington, D.C.: World Resources Institute, 1989.

Rothschild, Michael. *Bionomics: Economy as Ecosystem*. New York: Henry Holt, 1990.

Sarikaya, Mehmet, and Ilhan A. Aksay (eds.). *Biomimetics: Design and Processing of Materials*. New York: American Institute of Physics, 1995.

Soulé, Judith, and Jon K. Piper. *Farming in Nature's Image*. San Francisco: Island Press, 1992.

Swan, James A., and Roberta Swan. *Bound to the Earth: Creating a Working Partnership of Humanity and Nature*. New York: Avon, 1994.

Todd, Nancy. *Bioshelters, Ocean Arks, City Farming: Ecology as the Basis of Design*. San Francisco: Sierra Club Books, 1984.

Tributsch, Helmut. *How Life Learned to Live: Adaptation in Nature*. Cambridge, Mass.: MIT Press, 1982.

Viney, Christopher, Steven T. Case, and J. Herbert Waite. *Biomolecular Materials: Materials Research Society Symposium Proceedings*, Vol. 292. Pittsburgh, Pa.: Materials Research Society, 1993.

Vogel, Steven. *Life's Devices: The Physical World of Animals and Plants*. Princeton, N.J.: Princeton University Press, 1988.

Willis, Delta. *The Sand Dollar and the Slide Rule: Drawing Blueprints from Nature*. Reading, Mass.: Addison-Wesley, 1995.

Yeang, Ken. *Designing with Nature: The Ecological Basis for Architectural Design*. New York: McGraw-Hill, 1995.

INDEX

abalone shell, 97, 98–112, 142–143
Ableman, Michael, 11
Ableson, Philip H., 176
actin, 134
adhesives, byssus complex as biomodel
     for, 118–122, 124–129
Adleman, Leonard M., 231–235
Agricultural Revolution, 5
Agricultural Testament, An (Howard), 41
agriculture, 11–58
     animal diets as model for bioregional
          crop choices in, 160
     annuals vs. perennials in, 12, 14, 17,
          20, 25, 26–30, 46–47
     bioregional differences in, 35–36, 40–46
     government policies on, 16–17, 46–50
     industrialization of, 17–20, 53
     insecticide use in, 13, 14, 18–20, 47, 48–
          49, 179
     mature natural ecosystems as models
          for, 12–13, 21–27, 36–37, 40–41
     mixed species in, 14, 23–35
     of permaculture, 37–39
     soil damage from, 13–19, 25
     successional model of, 39–41
     sustainable energy for, 47, 50–53
     weather conditions and, 11–12
airplanes, development of, 8
Allenby, Braden M., 238, 241–243, 248,
          251–253, 254, 257, 258, 264–267,
          276–279, 281, 282
Allied Signal, 127–128
Altieri, Miguel, 34
aluminum industry, 95
amino acids, 102, 103, 109, 111
anaesthesia, biological consciousness
          models and, 220–224
Andow, Dave, 34
anhydrous ammonium fertilizer, 18
animals:
     agricultural, 44–45, 47–48, 51–52
     cooperative relationships between
          groups of, 258
     dietary discrimination of, 147–172, 178–
          184
     extinctions of, 139–140, 144–145, 146–
          147, 174
     habitat destruction for, 170
     toxic defenses used by, 179–181
anisotropy, 133
anting, 167
ants, 7
architecture, botanical models for, 6
Arizona, University of, 100, 219, 220
Arizona State University, 63
Army, U.S., spider silk protein research
          by, 133, 136–138
Arneson, Charles, 180
art, computer evolution of, 209–210
artificial intelligence, 195
artificial life, 202
AT&T, 238, 239, 268, 283
Atlantic Forest, 170
atmosphere, balance of gases in, 271–273
automobile industry, thin-film technology
          for use in, 113
Ayres, Robert U., 272–273

baboons, 151, 166
Bacon, Francis, 5
bacteria:
     in genetic engineering, 82, 106–108,
          208
     as light-sensitive digital switches, 213–
          219
     magnetotactic, 114–115
     in photosynthesis research, 70–74
bacteriorhodopsin (BR), 213–219, 229
Baum, Eric, 234
bears, medicinal plants used by, 167, 182
     grizzly, 8, 167
     polar, 6
Beck, Matthew, 169
Beethoven, Ludwig van, 185
Bell Laboratories, 239, 241
Bender, Marty, 52
Berenbaum, May R., 130, 179
Berger, Patricia, 171
Bergman, Morris, 236
Berish, Corey, 40
Berry, Thomas, 287–288, 295
Berry, Wendell, 16, 43, 53, 292, 293–294
BioComputing Group, 188
biodegradability, 126

Biological Components Corporation, 217
biomimicry:
  defined, xi, 2
  development of acceptance for, 4–6,
    287–295
  human capacity for, 295–296
  mechanical paradigm vs., 236
  potential applications for, 2–3, 6–7
Biophilia and the Conservation Ethic
  (Wilson), 285
Bioscience, 289–290
biosphere, chemical balance in, 271–273
biotechnology, genetic basis of, 107
Birge, Robert R., 216, 217–219
Black & Decker, 257
Body Shop, 257
bones:
  for dental implants, 144
  structural composition of, 101
Borges, Jorge Luis, 185–186
boundary conditions, 278–279
Bound to the Earth (Swan and Swan), 238
BR (bacteriorhodopsin), 213–219, 229
brain function:
  in accommodation to evolutionary
    changes, 200–201
  dietary selection and, 155, 158–159
  digital computers vs., 189–202
  electrochemical aspect of, 198–200
  at molecular level, 219–229
  of neural net parallelism, 196–198, 199
  quantum arguments for, 219–228
Brand, Stewart, 238–239
Broom, Jo Ann, 267–268
Brown, Robert, 104
Brownian motion, 104, 206, 212
Bugs in the System (Berenbaum), 130
bundleflower, Illinois, 27–29, 33–35
Burden, Jeremy, 34
Butz, Earl, 17
Byrne, Dick, 151
byssus complex, adhesive technologies
  based on, 118–122, 124–129

California, University of, at Santa Barbara,
  103, 111
Calvert, Paul, 100–101, 116–117, 125
cancer therapies, 146, 165, 172, 181
Canon, 257
carbon dioxide (CO_2), atmospheric levels
  of, 61, 271–273
carbon molecules, shape-based computing
  based on variety of, 194–195
Carson, Rachel, 244
cDNA (complementary DNA), 109–110,
  231
Celltak, 128
cellular automaton theory, 224

Center for Early Events in Photosynthesis,
  63–64
Center for Rural Affairs, 49, 50
ceramics production, 100
chameleons, 6
chemicals industry, toxic emissions from,
  95
chemivars, 177
Chesapeake Bay, mussels used to monitor
  metal residue in, 129
chimpanzees, self-medication by, 161–
  165, 168
Chiras, Daniel, 256, 271
chlorofluorocarbons (CFCs), atmospheric
  damage caused by, 122–123, 273,
  281
chloroplasts, 63, 65, 259
Ciamician, Giacomo, 62, 63, 69
cilia, 222
Clean Air Act (1989), 123
Clean Air Act (1990), 280
Clinton, Bill, 239, 280
coherent quantum states, 225–229
Colinvaux, Paul, 258, 261
Collaborative Research, 128
collagen, 119, 125
command-and-control laws, 244, 278
community-supported agriculture (CSA),
  56
complementary DNA (cDNA), 109–110,
  231
composite materials, 115, 116–117, 131,
  141–142, 265
compressional strength, 143
Computer, 230
computers:
  brain function vs., 186–187, 189–202
  evolution programs for, 209–211
  holographic memory in, 217–219
  miniaturization limit for, 191
computing, molecular, 202–237
  directed evolution for, 207–212, 229–
    230
  DNA processing approach to, 230–235
  fuzzy capability of, 206
  light-based digital operations for, 84–86,
    212–219
  microtubular model for, 219–230
  quantum theory in, 206
  shape-based interactions in, 193, 194–
    195, 203–207
  tactilizing processor posited for, 203–
    207, 208–209, 229
Comstock, Gary, 48–49
Congress, U.S., soil conservation efforts
  of, 16, 49
Connection Machine, 198
Conrad, Debby, 235
Conrad, Joseph, 57

Conrad, Michael, 187–213, 227, 234
  background of, 187, 188–189
  on brain function vs. silicon computers,
    189–202
  on molecular computing, 192–193, 198–
    200, 202–213, 219, 226, 229–230,
    233, 235–236, 237
  tactilizing processor conceived by, 203–
    207, 229
consciousness:
  biological structure at root of, 196–198,
    221–224
  quantum argument for, 225–228
Conservation Reserve Program (CRP), 46,
    47
Cooper, William, 248–253, 263, 264,
    276, 277, 278
cooperative systems, 258–260
CO₂ (carbon dioxide), atmospheric levels
    of, 61, 271–273
Cragg, Gordon, 146
Craighead, Frank, 183
Craighead, John, 183
Crawford, Michael, 154–155
crystallization:
  experimental technologies for, 111–118
  fourteen natural shapes of, 103
  protein templates for, 102–117
crystallography, 71–72, 131
CSA (community supported agriculture),
    56
Cullin, Dave, 217
Cultural Survival, 177
cuttlefish, 6
cytoskeletons, 219, 221–223, 229, 230

Daniels, Joe, 140–145
Dawkins, Richard, 210
Death of Nature, The (Merchant), 236, 241
Deisenhofer, Johann, 72
Déjà Shoe, 257
Delaware, University of, 118
dematerialization, 265
Deming, W. Edwards, 243, 264
dental surgery, bone growth and, 144
design:
  computer evolution programs for, 209–
    211
  environmental impact as factor in, 281–
    283
  role in sustainability movement, 282
diet, animal selection of, 147–172, 178–
    184
  in evolutionary process, 154–156, 158–
    159
  indigenous cultures' adoption of, 182–
    183
  for medicinal purposes, 161–169, 178
  olfactory senses used in, 156–157

  seasonal reproductive patterns linked to,
    169–172
  toxins avoided in, 147–151, 158–159,
    178, 183
directed evolution, 209
Dirks, Gary, 75
dirt eating, 153–154
DNA:
  complementarity of, 109–110, 231
  computer processing based on, 230–235
  in protein synthesis, 109–110, 238
dragonflies, 6
Drake, James, 31, 32
Dreams of Reason, The (Pagels), 185
Driving Force, The (Crawford and Marsh),
    154–155
duckweed, 59–60, 285–287
Duke University Primate Center, 146–
    147, 150
Du Pont, 262, 268
Dust Bowl, 13, 16, 17
Dwellings (Hogan), 285, 295

Earth in the Balance (Gore), 247
E. coli bacteria, protein synthesis with, 82,
    106–108, 110, 137–138, 208
Ecolyte, 87
Ehrenfeld, David, 289
Ehrlich, Paul, 148, 244
Einstein, Albert, 104, 247
Eisley, Loren, 8–9
Eisner, Thomas, 174, 175–176
electric cars, 113
Emperor's New Mind, The (Penrose), 226,
    228
emulsions, 123–124
endosymbiotic hypothesis, 258–259
energy production:
  in agriculture, 19–20, 50–53
  efficient use and, 260–263
  environmental damage caused by, 59, 61
  from fossil fuels, 61
  hydrogen as source of, 84
  photosynthesis model for, 59–63
  solar, 59–94
environmental degradation:
  in agriculture, 15–17, 19, 48–49
  as economic accounting factor, 278–
    281
  energy production as leader in, 59, 61
  from four materials industries, 95
  of ozone layer, 122–123
  preventive strategies for, 267–269
Environmental Protection Agency (EPA),
    Mussel Watch program of, 129
enzymes, 103, 104
  lock-and-key interactions of, 187
Estes, Richard, 161
ethnobotany, 176–177

evolution:
    capacity for change required in, 201, 202
    dietary choices and, 154–156, 158–159
    at molecular computing level, 202–203,
        207–212, 229–230
    sporadic progress of, 5
Ewel, John J., 12, 24, 40–41

fabbers, 115–117
family farms, decline of, 20
Farm Bill (1995), 49–50
Farming in Nature's Image (Piper), 15–16
Faulkner, D. John, 180, 181
feedback mechanisms, 274–275
fermentation, 157
ferritin, 115
fertility, seasonal, 170–172
fertilizers, 14, 18, 20, 44, 47
fiber production:
    by mussels, 124–125
    of spider silk, 132–136
Fischbach, Gerald D., 199
fish, 6
foam materials, 122–124
forestry, sustainable yields in, 271
forests:
    as models for agriculture, 38–43
    as models for business, 248–254
fossil fuels, solar energy in, 61
Franklin, Benjamin, 87
free-form manufacturing, 116
Freeman, David, 203
frogs, 6, 180
From Ecocities to Living Machines (Todd),
    39
Fukuoka, Masanobu, 36–37, 57
functional economy, service leases vs.
    product ownership in, 265–266
Furlong, Clement, 106–108, 110

Gaia hypothesis, 258, 294
Galef, Bennett G., Jr., 169
gamagrass, eastern, 28, 33–34
Gathering the Desert (Nabhan), 43
gel electrolysis, 108–109
General Motors, 113, 239
genetic algorithms, 211
genetic engineering, 5, 107–109
Geography of Childhood, The (Nabhan and
    Trimble), 288
Gifford, David, 230, 233
Glander, Kenneth, 146–147, 149, 150,
        153, 156, 158, 168–169, 171, 178,
        183
glass:
    energy employed in production of,
        95
    silane primers used on, 121
glass-transition temperatures, 132

Global Business Network, 239
Goldberg, Ina, 128
Goodall, Jane, 163, 164, 168
Gore, Albert, Jr., 247
Graedel, Thomas, 266, 282–284
Grassland, The (Manning), 15, 18
Great Plow-up, 16
Greenler, Robert, 131
Green Revolution, 17, 160
groundwater, contamination of, 19
Guillet, James, 86–94
Gust, J. Devens, Jr., 63–64, 69–70, 74–
    75, 76–82, 85–86, 87

hailstorms, 12
Halobacterium halobium, 213–214, 215,
    216–217
Hameroff, Stuart, 219–230, 235
Hamiltonian path problem, 231–233
Hardy, Thomas, 293
Hart, Robert, 41
Hassebrook, Chuck, 49
Havel, Václav, 1
Hawken, Paul, 239, 280
Heat-Moon, William Least, 54
heavy metals, 128–129
Hebb, Donald O., 197, 222
Hegel, G.W.F., 228
Henderson, Hazel, 281
Hogan, Linda, 285, 295
holistic medicine, 4
Holland, John, 211
holographic memory, 217–219
Home Economics (Berry), 292
Hong, Felix, 207, 213–215
horsehair, 144
Howard, Sir Alfred, 41
How to Survive on Land and Sea
    (Craighead and Craighead), 183
Huaorani Indians, 1, 3
Huber, Robert, 72
Huffman, Michael, 146, 161–163, 164,
    168, 182
Hughes Aircraft Corporation, 216
Humbert, Rich, 98–99, 101–102, 105,
    106, 108–110
hummingbirds, 7
Hunter, Emily, 54–55
hybrid cultivation, 17
hydrogen gas, as fuel source, 84

immune system, 204
indigenous cultures:
    biomimicry in dietary choices of, 182–
        183
    botanical lore from, 176–177
    disappearance of, 177
    nature respected by, 1, 3, 9, 11, 293,
        294–295, 297

industrial ecology:
  development of, 239–248, 283–284
  goal of, 242
  mechanisms for economic shift to, 277–
    284
  natural systems models for, 248–254
  ten principles for, 253–277
industrialization:
  of agriculture, 17–20, 53
  environmental damage linked with
    energy sector for, 59
Industrial Revolution, 5
  limits on, 238, 242, 249
  mass production and, 264
  nature exploited in, 2, 238
information:
  from environmental feedback
    mechanisms, 274–275
  human capacity for handling of, 294
Ingenhousz, Jan, 60
insecticides:
  agricultural use of, 13, 14, 18–20, 47,
    48–49
  plants as source of, 179
insulin, 82, 103, 107, 208
International Society for Molecular
    Electronics and Biocomputing, 206–
    207
Ireland, C. M., 180
iron-oxide crystals, 114, 215–216

Jackson, Dana, 22
Jackson, Laura, 28
Jackson, Wes, 9, 11, 12, 14, 16, 28, 293
  background of, 21–22
  on farming community, 20, 53–54, 55
  Natural Systems Agriculture promoted
    by, 21, 35–36, 37, 46, 48, 49–50,
    56, 57
  on perennial polycultures, 26–27, 30,
    44, 51, 58
Jacobson's organs, 156
Janzen, Dan, 175
Javanese agriculture, 39–40
Jefferson, Thomas, 20
Johns, Timothy, 154
Johnsongrass, 29
Joint Program on Drug Discovery,
    Biodiversity Conservation and
    Economic Growth, 174
Joyce, Gerald, 209
Jung, Carl, 228
just-in-time manufacturing, 269

Kaplan, David L., 133, 136–138
Kaufmann, Stuart, 24
Kelly, Kevin, 186, 196
keratin, 140, 141–143, 144

Kevlar, 132, 135, 137
Koyukons, 294

Land Institute, The, 11, 20–21, 22–36, 46–
    56
Langmuir-Blodgett (L-B) film, crystal
    growth on, 111–112
Laoch, Paul, 75
lasers, 88, 225
Laudise, Bob, 239–240, 243–244, 246–
    247, 257, 266, 268, 278, 284
Lawrence, Rick, 216
lemurs, 146–147, 150
Letters to the Earth (Twain), 8
Levy, Steven, 234–235
Lewis, Randolph V., 127, 129, 133, 135,
    138
Liberman, E. A., 198
"Library of Babel, The" (Borges), 185–
    186
Life Cycle Analysis (LCA), 283
light:
  digital information processing based on,
    85–86, 212–219
  laser, 88, 225
limits, power of, 7–8
Lipkin, Richard, 132
liquid crystal, 131, 133
local economic self-reliance, 276–277
Lovins, Amory, 262, 269
Lyons, Oren, 288

McChesney, Charles, 173
McKey, Doyle, 151
McKibben, Bill, 8
Magainin Inc., 180
magnetic media, iron-oxide crystals in,
    114, 215–216
magnetotactic bacteria, 114–115
Mann, Stephen, 114–115
Manning, Richard, 15, 18, 20
Margulis, Lynn, 258–259
marine life, pharmaceutical research on,
    173, 180–181
Marriott, Bernadette, 152–154, 160
Marsh, David, 154–155
material resources, minimal use of, 264–
    267
materials exchange brokerages, 274
Materials Research Society (MRS), 95–98,
    112
materials science, biomimicry in:
  abalone nacre and, 99–112
  biodegradability models for, 126
  in composite products, 115, 116–117
  four principles of natural materials and,
    96, 117
  with inorganic structures, 98–118

materials science (continued)
  life-friendly manufacturing processes
    from, 96, 97
  mussel adhesion and, 118–129
  ordered hierarchical structures in, 96,
    99–100
  with organic models, 117–129
  protein templates in, 96, 102–117
  rhinoceros horn and, 97, 140–145
materials science, (continued)
  self-assembly process of, 96, 101–104
  spider silk and, 129–139
Matfield Green Project, 54–55
Mathis, Paul, 77
medicines:
  in animal diets, 161–169, 178–184
  directed evolution of, 209
  of indigenous cultures, 176–177, 182
  plant sources for, 147, 161–169, 172–
    179, 182–183
  toxic defenses as source for, 179–181
membrane potential, 67–68, 86
Mendel, Gregor, 31, 209
Merchant, Carolyn, 236, 241
metal production, 95
metal residues, mussels used in
    monitoring of, 128–129
Michel, Hartmut, 71–72
microtubules, 219, 221–230
Milton, Katherine, 150–151
Mission to Planet Earth, 273
Moi (Huaorani leader), 1, 9
Mollison, Bill, 37–39, 44, 57
monkeys:
  black colobus, 151
  howler, 150–151, 158, 166–167, 171
  red colobus, 168
  rhesus, 153, 154
Montreal Protocol on Substances that
    Deplete the Ozone Layer, 123
Moore, Ana, 69, 74, 76–78, 79, 82, 83
Moore, Thomas A., 64–69, 74–78, 82, 83,
    85–86, 87, 90
Morse, Daniel, 103, 111–112
MRS (Materials Research Society), 95–98,
    112
muriqui, 170–171
mussel, blue, 118–129
mussels, adhesive functions of, 97, 118–
    122, 124–129
Mussel Watch, 129
mutagenesis, 72

Nabhan, Gary Paul, 12, 43, 288
nacre, 99–112
National Academy of Sciences, 152
National Biological Survey, 292
National Cancer Institute, 175

National Cancer Laboratory, 146
National Institutes of Health, 165, 173
National Science Foundation, 64
Natural Product Sciences, 179
nature:
  humanity's place in, 1–2, 3; 8–9, 11,
    241–242, 287–295
  indigenous peoples' respect for, 1, 3, 9,
    11, 293, 294–295, 297
  nine basic laws of, 7
  reductionist views of, 236–237
Naval Research Center, 216–217
Nelson, Richard, 294
neural nets, 196–198
neurotransmitters, 194, 199, 222
New Alchemy Institute, 39
Newsweek, 234
Newton, Isaac, 226
New York Botanical Garden, 173–174
nitrates, 19
Nixon, Richard, 17

Odum, Howard T., 251
Office of Naval Research, 106
Old Dominion University, 140
olfactory senses, 156–157
oligio machines, 208
One Straw Revolution (Fukuoka), 37
opportunistic systems, 249–250
ozone layer, CFC destruction of, 122–
    123, 281

Pacific Northwest Laboratories, 112
packaging, waste reduction in, 256–257
Pagels, Heinz, 185
paint, primers for, 119–120, 121
Papago agriculture, 43
paper industry, 95
parallel processing, 190, 191, 196–198,
    231, 234
Patagonia, 257–258
pattern recognition, 190, 195, 203
PCBs (polychlorinated biphenyls), 91–92
Penrose, Roger, 220, 226, 227, 228
pesticides, 13, 14, 18–20, 47, 48–49, 179
petroleum products:
  in agriculture, 18–20
  photosynthesis as source of, 61
Pfizer, Inc., 173–174, 179
pharmacology, see medicines
phase inversion, 124
pheromones, 130
Phillips-Conroy, Jane, 166
photosynthesis, 60–94
  artificial models of, 62–83, 86–94
  defined, 60
  digital technology and, 84–86
  early studies on, 60

efficiency of, 59, 79, 80–81, 90, 260
  molecular process of, 62–63, 215, 217
  photozyme creation and, 86–93
photovoltaics (PVs), 62
photozyme, 86–93
Picattiny Arsenal, 123, 124
Pimentel, David, 19
Pimm, Stuart, 31, 32
Piper, Jon, 15–16, 23–26, 29–35, 46, 47,
  51, 56
plants:
  animal fertility and, 170–172
  annual vs. perennial, 12, 14, 17, 20, 25,
    26–30, 46–47, 249–251
  extinctions of, 147
  medicinal use of, 161–169
  photosynthesis of, 59–94, 215, 217
  poisonous, 147–149, 150
  see also agriculture
plastics industry:
  solid foam production in, 122–124
  toxic emissions produced by, 95
plowing, damage from, 13, 16
polarizing light microscope, 133, 141
pollution, *see* environmental degradation
polychlorinated biphenyls (PCBs), 91–92
polystyrene, 113, 123–124
population levels, 240, 249, 292
prairie ecosystems, 12
  four classic plant types of, 26
  lack of soil damage in, 14–19, 25
  sustainable agriculture modeled on, 20–
    35, 44–53
pregnancy, dietary preferences in, 157
Priestley, Joseph, 60
primates, dietary choices of, 147, 149–172
  discrimination abilities for, 156–159,
    169
  medicinal plants in, 161–169
  nutritional values vs. toxic levels in, 149–
    151
  sampling behaviors in, 158
  for seasonal fertility, 169–172
  soil ingestion and, 153–154
primers, 119–121
productivity, optimal vs. maximal, 263–
  264
Profet, Margie, 157
profitability, environmental costs
  considered in, 278–281
proteins:
  digestion of, 148–149, 170
  self-assembly of, 103–104
  sequencing research on, 105–106, 107,
    108–109, 111
  as structural templates, 102–105, 110–
    117
  synthesis of, 106–108, 109–111, 208

  see also computing, molecular
PVs (photovoltaics), 62

quantum theory:
  brain function and, 225–228
  parallel scanning and, 206, 226–227

radar, 6
Rámon y Cajal, Santiago, 185
Reenchantment of the World, The
  (Bergman), 236
Reidel, Evie Mae, 55
resource levels, maintenance of, 269–271
rhinoceros, 139–145
rice, 17
Richter, Curt P., 152
Ride, Sally, 273
Rieke, Peter C., 112–114
Risch, Steve, 34
Rittmann, Stephanie, 45–46
Roberts, Pat, 49
Rodale family, 44
Roddick, Anita, 257
Rodriguez, Eloy, 146, 165, 166, 182
Ryan, Christopher, 282, 284

Sachs, Aaron, 95
salad dressing, emulsified, 123–124
Sales, Kirkpatrick, 277
SAMs (self-assembled monolayers), crystal
  growth on, 112–114
Sapolsky, Robert M., 182
Sarikaya, Mehmet, 95, 99, 101, 105
Sauther, Michelle L., 159
Schapiro, Mark, 19
Schrödinger, Erwin, 227
Schultes, Richard Evans, 177
Science, 62, 176, 230, 233, 234
Science News, 132
Sciences, The, 177, 182
Scientific American, 199
Scientific Revolution, 5, 241
Scott, Thomas, 157
Scripps Research Institution, 209
SCS (Soil Conservation Service), 16–17
Seifu, Mohamedi, 161
self-assembled monolayers (SAMs), crystal
  growth on, 112–114
Shadows of the Mind (Penrose), 220, 228
Shaman Pharmaceuticals, 177
shark, dogfish, 180
Sigsted, Shawn, 167
silanes, 121
silk, 97, 127, 129–139
silkworms, 131
smart materials, 6
Smith, Adam, 279
Smith, Ben, 21–22

Smith, J. Russell, 12–13, 41–43
soap, chemistry of, 90–91
sodbusters, 16
soil:
    agricultural impact on, 14–19, 25
    dietary ingestion of, 153–154
Soil Conservation Service (SCS), 16–17
solar energy:
    man-made efforts in usage of, 59, 81–82
    in photosynthesis, 59–94
solaron beads, 92
solid foam, 122–124
sorghum, milo grain, 29
Spanish dancer, 180–181
spider, golden orb weaver, 129
spider silk, 97, 129–139
    production process for, 132–136
    protein synthesis research based on, 136–138
Stealth bomber, 142
steel industry, 95
Stegner, Wallace, 46
Strange, Marty, 50
Strier, Karen B., 146, 163–164, 170–171, 178, 182
Stuckey, Galen, 111, 112
Styrofoam, 122, 123–124
Sunshine Farm project, 51–53, 55
superconductivity, 22
superimposition of states, 226–227, 228
Swan, James A., 238
Swan, Roberta, 238
symbiosis, 258–259
Syracuse University, 216

tactilizing processors, 203–207, 208–209, 229
take-back laws, 256–257, 278
Tallgrass Prairie Producers, 55
Tate, Ann, 216–217, 235
teeth, crystalline structure of, 101
tendons, structure of, 99–100, 119, 125
Tennyson, Alfred, Lord, 258
termites, 6
thin-film technology, 112–114
3M, 267–268
Tibbs, Hardin B., 239, 241, 246, 247, 254
Todd, John, 39–40
Todd, Nancy, 39
torsional strength, 143
Total Quality Management (TQM), 243
toxins:
    in animal defense systems, 179–181
    dietary avoidance of, 147–149, 150–151, 158–159, 178, 183
    in self-medication, 168–169

Tree Crops (Smith), 41–43
Tributsch, Helmut, 200–201
Trimble, Stephen, 288
Twain, Mark, 8
Type I systems, 249–253
Type II systems, 250
Type III systems, 250–253

Ubby, Russell, 56
Uenohara, Michiyuki, 259
Ulmer, Kevin, 231
Ultimate Computing (Hameroff), 230
Understanding Chimpanzees (Goodall and Wrangham), 164–165
United Nations Conference on Environment and Development, 174

Van Orden, Ann, 140–145
Velcro, 4
Vermeer, Donald E., 154, 183
vervets, 151
Viney, Christopher, 129–136, 138–139, 141
Voisin, André, 45
von Neumann, John, 192, 196, 224

Waite, J. Herbert, 118–129
Washington, University of, 99, 106, 129
waste:
    annual tonnage of, 240–241
    industrial output of, 95, 243, 246
    public awareness on, 244–246
    as resource, 254–258, 274
water, chemistry in, 90–94
Waterman, Peter, 151
water supply, pesticide residues in, 19
Watt, Rich, 222–223
wave function, quantum, 227, 228
wheat cultivation, 13, 16
Whiten, Andrew, 151
wildrye, mammoth, 28, 33
Wilson, E. O., 285, 289–290, 296–297
With Bitter Herbs They Shall Eat It (Johns), 154
Woodbury, Neal, 70–74, 80, 83, 88, 91
Worman, Jack, 53
Wrangham, Richard, 146, 151, 156, 159, 162, 163–165, 166, 168, 178, 182
Wyoming, University of, 127, 133, 138

X-ray crystallography, 71

Young, John E., 95

Zasloff, Michael, 180
zoopharmacognosy, 166, 182